Sound and Safe

Sound and Safe

A History of Listening Behind the Wheel

Karin Bijsterveld,

Eefje Cleophas,

Stefan Krebs,

and

Gijs Mom

Oxford University Press is a department of the University of Oxford.
It furthers the University's objective of excellence in research, scholarship,
and education by publishing worldwide.

Oxford New York
Auckland Cape Town Dar es Salaam Hong Kong Karachi
Kuala Lumpur Madrid Melbourne Mexico City Nairobi
New Delhi Shanghai Taipei Toronto

With offices in
Argentina Austria Brazil Chile Czech Republic France Greece
Guatemala Hungary Italy Japan Poland Portugal Singapore
South Korea Switzerland Thailand Turkey Ukraine Vietnam

Published in the United States of America by
Oxford University Press
198 Madison Avenue, New York, NY 10016

Library of Congress Cataloging-in-Publication Data

Bijsterveld, Karin, 1961–
Sound and safe : a history of listening behind the wheel / Karin Bijsterveld, Eefje Cleophas,
Stefan Krebs and Gijs Mom.
pages cm
Includes bibliographical references and index.
ISBN 978–0–19–992569–8 (cloth : acid-free paper) 1. Automobiles—Noise—
History. 2. Automobiles—Design and construction—History. 3. Automobiles—Audio
equipment—History. 4. Automobiles—Safety—History. 5. Sound—History. 6. Auditory
perception—History. 7. Privacy—History. 8. Automobile driving—Social aspects—
History. I. Cleophas, Eefje, 1983– II. Krebs, Stefan, 1973– III. Mom, Gijs, 1949– IV. Title.
TL246.B55 2013
629.2′31—dc23
2013009907

9 8 7 6 5 4 3 2 1
Printed in the United States of America
on acid-free paper

CONTENTS

ACKNOWLEDGMENTS

This book is about seeking auditory privacy on public roads, but the book itself has not exactly been made in private. It started, as do all books, with bringing a few ideas together, but took off with the help of public funding, granted by the Netherlands Organization for Scientific Research (NWO). We did not work on our book in silence either. As soon as we had funding, we intended to integrate our research into a monograph at the end of the project. But we could only meet this goal by presenting our initial ideas to audiences of students and scholars in history, music, sound studies, media studies, and science and technology studies—most of whom were also drivers. These students and scholars always commented on our work from both their academic and everyday experiences, and our project benefited from these responses. The same has been true for the publications that resulted from our research: articles and chapters in edited volumes greatly improved by comments given by colleagues, anonymous peer reviewers, and editors.

Because we published such articles and chapters with the structure of our book in mind, we have been able to draw on these earlier publications when assembling it. At the same time, we shortened, integrated, recombined, and extended these articles with new examples, illustrations, sections, and three novel chapters. We are grateful for the courtesy of the publishers who granted us permission to reuse and rework sections of the following articles. For chapter 2: S. Krebs (2011), "The French Quest for the Silent Car Body: Technology, Comfort and Distinction in the Interwar Period," *Transfers: Interdisciplinary Journal of Mobility Studies* 1(3), 64–89 (courtesy Berghahn), and one section from G. P. A. Mom (2011), "Encapsulating Culture: European Car Travel, 1900–1940," *Journal of Tourism History* 3(3), 289–307 (courtesy Taylor and Francis). For chapter 3: S. Krebs (2012), "'Sobbing, Whining, Rumbling': Listening to Automobiles as Social Practice," in T. Pinch and K. Bijsterveld, eds., *The Oxford Handbook of Sound*

Studies, pp. 79–101 (Oxford: Oxford University Press) (courtesy Oxford University Press), and K. Bijsterveld (2010), "Acoustic Cocooning: How the Car Became a Place to Unwind," *Senses & Society* 5(2), 189–211 (courtesy Berg). For chapter 5: E. Cleophas and K. Bijsterveld (2012), "Selling Sound: Testing, Designing and Marketing Sound in the European Car Industry," in T. Pinch and K. Bijsterveld, eds., *The Oxford Handbook of Sound Studies*, pp. 102–124 (Oxford: Oxford University Press) (courtesy Oxford University Press), and a few remarks from S. Krebs (2012), "Standardizing Car Sound—Integrating Europe? International Traffic Noise Abatement and the Emergence of a European Car Identity, 1950–1975," *History and Technology* 28(1), 25–47 (courtesy Taylor and Francis).

Parts of chapter 4 have been presented as (unpublished) papers by Karin Bijsterveld, one of which was "'Like a Boxed Calf in a Traffic Gutter': The Roadside Noise Barrier in the Netherlands, 1970–2010," at the SHOT Annual Meeting, Cleveland, Ohio, November 3–6, 2011, and section 7 of chapter 2 draws on a paper under review: G. Mom, "Orchestrating Car Technology: Noise, Comfort, and the Construction of the American Closed Automobile, 1917–1940." Finally, we would like to mention that section 3 of chapter 5 builds on the second chapter of Eefje Cleophas's PhD dissertation in the making.

In the early stages of our project, Kristin Vetter and Fleur Fragola did some of the interviews for chapter 5, for which we are still grateful. We would also like to thank Ton Brouwers, Margaret Meredith, and Petra van der Jeught for correcting our English; the Faculty of Arts and Social Sciences at Maastricht University for funding the costs of correction; and Lidwien Hollanders for doing preparatory work for the reference list. Moreover, we have had excellent help from Marianka Louwers (Philips Company Archives Eindhoven), Beate Kuhn and Dietrich Kuhlgatz (Bosch Archives Stuttgart), Andy O'Dwyer (BBC Archives), and Mark Patrick (National Automotive History Collection, Detroit, MI). Our research would simply have been impossible without the willingness of those interviewed (see the list of interviewees), and the cooperation of Foort de Roo, Leif Nielsen, and Lotte Sørensen, who shared ISO working group archives with us. Finally, we would like to thank Norman Hirschy, editor at Oxford University Press, for all of his wise advice, and the anonymous reviewers of our book proposal for their insightful comments.

Sound and Safe

CHAPTER 1

Driving in Control

SECLUDED AND CONNECTED

Each and every morning, millions of drivers get into their cars. They close the door, insert the ignition key, and start the engine. From the moment the engine is humming and the car is ready to leave its parking space, the driver is both secluded from and thoroughly connected to the surrounding world.

What is the *seclusion* we are talking about? In more and more Western countries, noise barriers block the driver's view of the landscape. Instructions on speed and direction displayed on the matrix panels above the road lead the motorist through an almost sealed corridor of networked highways. Numerous technologies help the driver to experience the interior of the car as a private, secure, and self-enclosed space. The heater and air conditioning control the interior temperature. The engine is sufficiently quiet to enable the driver to listen to a favorite radio station or iPod song. The look, the sound, the smell, and even the feel of the car's interior have been designed carefully, tailored and targeted to the type of consumer behind the wheel. It is a design that enables the driver to feel, in our twist of the common phrase, sound and safe. At the very same time that the motorist experiences this seclusion, however, navigation technology, a mobile phone, and traffic information on the radio, *connect* the driver to what is happening outside the car. These services help motorists to find their way and keep moving—even though moving is exactly the thing they may be least sure about, notably in urban areas.

This book is about the process by which, over the past several decades, engineers and designers fashioned cars into mobile listening booths, in particular how the car's *sonic* design is interrelated with the history of highways. Although the limited-access road and its traffic jams confine the putative freedom of individual drivers and funnel them through their journey, the car's encapsulating sonic features and audio systems bring auditory privacy, musical entertainment, and traffic information that compensate for some of the control drivers have lost along the way. To put it differently: the sonic capsule enables the phenomenon of acoustic cocooning, or the driver's ability to relax by controlling the car's interior acoustic environment, in traffic situations that allow only limited freedom in other aspects of driving. It is important to note from the outset, however, that the histories of the highway and the mobile listening booth became intertwined only gradually. There is, in other words, no inherent logic to automobiles evolving into the sonic capsules they are today. The early automobile, in fact, was a highly noisy vehicle, often desirably so, underscoring the very *unlikelihood* of the development we address here—of cars evolving into mobile listening booths. And yet this is precisely what happened. This book will trace how, exactly, this metamorphosis occurred.

It does so by unraveling the history of cars on the freeway corridors of Western countries through the medium of sound, and with an ear for the rapidly changing mobility and auditory cultures from the 1920s until today. Our story will be full of engines, bodies, doors, wipers, blinkers, tires, radios and transistor sets, sound insulation technologies, noise barriers, sound-deadening asphalt, and sound simulation vehicles. How did these artifacts sound, or make things sound? Who listened to them and in which ways? Which experts intervened in these sounds, for what reasons and with what results? Herein is a history of listening behind the wheel—listening by drivers as well as by journalists who tested cars and engineers who tweaked automotive sound.

Our empirical research is focused on Western Europe and North America. With respect to several of our topics we started in Europe, yet grasped every opportunity to draw comparisons with developments in the United States, if only because our protagonists among car manufacturers, automotive engineers, and car consumer organizations were happy to do so. Americans dominated the automotive industry in the interwar years. And since we have centered our empirical work for each period we studied on those countries in which particular manufacturing industries, testing laboratories, consumer associations, and traffic and road authorities took the lead in the sonic issues that interested us, the United States figures prominently in the chapter about the 1920s and 1930s.

Our approach also means that we had a special ear for Germany when studying sound design and testing in the automotive industry after World War II, since German car manufacturers became increasingly important in this period. The early professionalization of car mechanics in Germany also made this country a relevant site of study. In addition, we focus on the Netherlands when narrating the history of car radio in Europe, partly because Philips, a major

Dutch electronics company, was an early and significant player in the business of making and selling auto radio. To mention yet another example, we head to France to learn more about a unique approach to silencing car bodies. In none of the chapters, however, did we limit our focus to one country. We have always widened our geographical scope beyond a single country and, as we have said, often draw in relevant developments in the United States. We thus arrive at many destinations, much like the drivers we are interested in.

WHY *ACOUSTIC* COCOONING AND THE *SOUND* OF CARS?

To be sure, we are not the first to point to the secluded character of a car's interior. The rise of the car as a mobile living room and its unique position among transport technologies as an artifact that provides both mobility and a privatized, intimate space has been underlined again and again. In fact, several scholars consider this unique combination of affordances to be an important explanation for the car's long-term success (Urry 2000a; and see Shove 1998 for an overview of explanations).

Manufacturers initially advertised the car as an adventure machine for sportive men, and then, from the 1920s onward, gradually redirected their marketing strategies to traveling businessmen and touring middle-class families, redrafting their sales pitches to underline the automobile's easy use and interior comfort (Flink 1992; Koshar 2002; Mom et al. 2008). It even could serve, as automotive historian Michael Berger has written, "as a parlor or bedroom on wheels" (2001: xxi). And an increasing number of cultural histories of automobility have clarified—by analyzing fiction such as road novels and road movies and nonfiction travelogues—how drivers and passengers used the intimacy of a car to get to know the world, each other, and themselves. Symbolically, traveling by car often stands for a journey into one's inner self (Cohan and Hark 1997).

Only recently have psychologists and sociologists begun to stress the role of sound in this search for mobile seclusion. According to transport psychologists, today many people view their vehicle as a place to "unwind," as a site where "neither colleagues nor the hustle and bustle of family life" will bother them (van Paassen 2004: 1).[1] Driving home from work, listening to the radio or talking on the phone, commuters enjoy a "peaceful moment" for themselves right before their domestic rush hour starts with "hungry children" or other demands claiming their attention (Brinks

1. In this book, all translations of quotes from German, French, and Dutch sources have been made by the authors.

2011: 12). Drivers' appreciation of a "social vacuum" has been put forward as an explanation of the failure of carpooling (Wildervanck and Tertoolen 1997: 20; van Lieshout 2006, 2007).[2] After all, to those who turn off their cell phone or use it only to place calls, the car may well be the last bastion of privacy—one that most drivers are unlikely to give up anytime soon.

Our sense of sound contributes considerably to this experience of privacy. Since the late 1990s, car manufacturers have increasingly advertised their cars' interior tranquility and subtle sound design. Television commercials by Toyota, Mercedes-Benz, and Volkswagen have stressed that their vehicles provide the quietude that is absent outside.[3] Many newspaper and magazine ads similarly address the car's interior sound. The Chrysler Voyager, so its manufacturer claims, has a "whisper-silent" interior.[4] In the Ford Focus all that "you touch" makes "the sound it is supposed to make" (advertisement, Ford 2005). "An individual receives 469,082 sound stimuli per minute," BMW says. "It starts with the alarm clock. And it usually ends with 'goodnight' (or a snore). In between your ears are on overtime. Every so often, then, tranquility is the right thing to seek. For instance, in a diesel of BMW" (advertisement, BMW 1999).

One cannot overlook the irony in the fact that marketers present the automobile as a place to unwind in peace and quiet—at least not from the perspective of the nonmotorist. Surveys show that residents of EU urban areas consider road traffic the most annoying source of noise.[5] But *inside* the car, tranquility is within easier reach than it has ever been, even if we acknowledge the differences between more and less expensive cars and take into account manufacturers' tendency to overstate the quality of their cars' acoustic interiors. An advertisement by Acura (figure 1.1) portrays its car side by side with a soundless musical score, a score with empty staffs. The accompanying text says:

> Most speakers only create sound. Ours, on the other hand, can also take it away. Microphones inside the cabin constantly monitor unwanted engine noise.

2. The journalists writing about these topics (Daphne van Paassen, Mirjam Brinks, and Marcel van Lieshout) notably refer to the research done and claims made by the Dutch transport psychologists Karel Brookhuis and Cees Wildervanck. See also http://www.politie.nl/Brabant-Zuid-Oost/Images/dit4webversie%2067_tcm27-333145.pdf (accessed May 10, 2012).

3. "Business Travel" (1990), commercial, Mercedes-Benz, videotape. Stuttgart: DaimlerChrysler AG. "Blind Brothers Three" (1999), commercial, Volkswagen Passat, videotape. Almere: TeamPlayers. For a more detailed discussion of these and similar commercials, see chapter 5.

4. "Let your car do the talking." Magazine ad for the Chrysler Voyager, published between 2002 and 2007.

5. European Local Transport Information Service, at http://www.eltis.org/show_news.phtml?newsid=1035&mainID=&Id=all (accessed April 21, 2009).

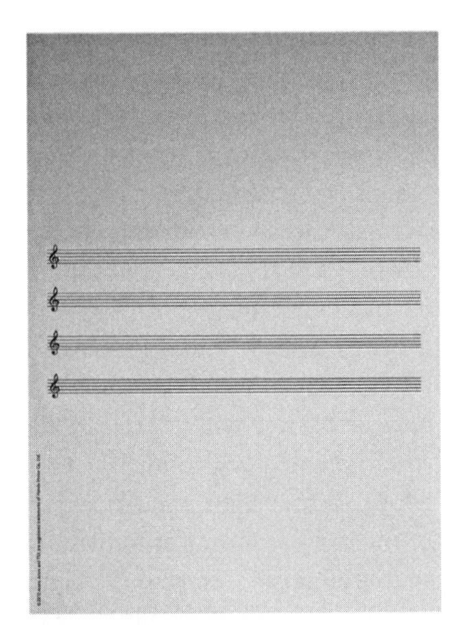

Figure 1.1
"Our speakers can create an interesting sound: Silence." Advertisement, Honda Acura 2010.
Source: http://1.bp.blogspot.com/_m5pQcjFdzMg/TLdZFWjmS2I/AAAAAAAABDU/6bIvKb1phW0/s1600/acura-ad.jpg, retrieved May 7, 2012.
Courtesy: American Honda Motor Co., Inc.

> When noise is detected, opposing frequencies are broadcast through the speakers to eliminate it, literally fighting sound with sound. The result is dramatically reduced engine noise for a quieter, more comfortable cabin.[6]

Quietness is not a necessary condition for finding peace in cars, however, as long as drivers *feel* themselves to be in control of their acoustic environment. A sense of privacy, social psychologist Irwin Altman has claimed, is dependent not so much on a person's isolation as on the control the individual has over access to his or her domain (Altman 1976). This psychological tendency also applies to auditory privacy in cars. As media sociologist Michael Bull has shown, listening to radio, CDs, or iPods offers drivers the feeling they are in charge of their journey. They may turn an unpredictable trip into something more routine and comfortable by tuning into the same radio program over and over again. Or they do the opposite, turning a boring commute into an interesting one by listening to the audiobook

6. Advertisement Acura (2010), at http://1.bp.blogspot.com/_m5pQcjFdzMg/TLdZFWjmS2I/AAAAAAAABDU/6bIvKb1phW0/s1600/acura-ad.jpg (accessed May 7, 2012).

version of the novel they always wanted to read but never got around to (Bull 2001, 2003, 2004). Drivers can play their music loudly and sing out of tune without bothering others—well, up to a certain decibel level. "Great acoustics" is a Dutch writer's recent assessment of his car's interior (Dijkshoorn 2011: 37). He and other drivers sing along with music on the radio or iPod playlists in a space that seems acoustically sealed off from the outside world. This isolation renders that world a sort of movie, not quite completely real (Bull 2003, 2004; see also Stockfelt 1994).

Today driving itself—given the traffic jams, speed limits, prescribed routes, and roadside noise barriers we have already referred to—does not match the ideal of freedom the car was once associated with (Shove 1998; Urry 1999, 2000b; Seiler 2008). We all tend to roll along in highly standardized "passages" from A to B (Peters 2006). At first sight, then, it seems we can explain the paradoxical position of the car—as means of individual mobility *and* object of traffic control, as source of noise nuisance *and* realm of tranquility—psychologically. Precisely because we are so readily mobile, so restlessly reachable and disconsolately constricted, the vehicle in which we travel has become the preeminent place to relax and come to our senses. Scholars can only rely on such a psychological portrait of the driver, however, when they take it as a given that drivers can listen to whatever they enjoy. Yet it is extremely odd that designers, manufacturers, and consumers have turned a device propelled by a noisy internal combustion engine into a listening booth.

How, then, did the automobile develop into what Michael Bull has called a "sonic envelope" (2004: 247), sociologist John Urry a "mobile capsule" (1999: 9), and sound theorist Brandon LaBelle a "sonic bubble" (2008: 193)? How did the car end up as a space for "acoustic cocooning," that is: a domain in which people experience privacy and relaxation because the interior acoustics of cars are pleasant and controllable? Since when, for instance, do we have car radios? How did the interest in the exterior and interior sound of cars arise? Who was listening to cars, and how did these people respond to what they heard?

It is true that other means of passenger transport, such as trains and airplanes, have also become quieter places over time. Just like cars, they are subject to noise regulation and control. But they do not offer the same level of auditory privacy as cars. Many people use headphones to listen to their music while traveling by aircraft, and it certainly helps to create some auditory seclusion. But singing along would be frowned upon by fellow passengers. The information train and plane travelers receive over the radio does not provide them freedom to choose their way. An aircraft passenger may feel alarmed by a sudden engine sound, but cannot intervene in what causes it. Moreover, car manufacturers have developed a personalization of interior vehicle sound that is absent from trains and airplanes. Nowadays, most individuals in the West encounter automobiles far more frequently than trains and airplanes. The conclusions we draw from studying the car have thus more to reveal about the relationships between sound and society than attention to any other form of passenger transport.

We acknowledge that the process of encapsulating the driver is linked to many technologies besides merely sonic ones. The car's climate control, cruise control, automatic transmission, information systems, and sumptuous seats all contribute to its enveloping character (Urry 2000a; Sheller 2004). On the other hand, not all of a car's encapsulating features are under the control of, or are chosen by, its driver. "Prosthetization" and computerization have reduced drivers' direct, sensory experience of the engine, brakes, and steering system (Mom 1997, 2008a; Sheller 2004). If this buffering contributes to encapsulating drivers, it does not enhance their *options* for cocooning. But many of the sound technologies we discuss in this book *do* enhance the driver's options, and it is exactly for this reason that *acoustic* cocooning intrigues us. That is why we will use "acoustic cocooning" and "mobile listening booth" more often than the related terms "sonic envelope," "sonic bubble," and "mobile capsule." Even though the word "cocooning" connotes seclusion, in our use it refers to control over a sonic environment—feeling sound and safe—which implies seeking both isolation and connection through sound.

This book thus narrates the histories of *shielding* the car's interior acoustic space from outside noises, and of *filling* this space with the sound of audio equipment, the auditory signals of seatbelts and blinkers, the sonic expressions of wipers and windows, and so forth. Moreover, we distinguish *acoustical* from *metaphorical* shielding and filling. While acoustic shielding and filling refer to sound insulation and sound production respectively, filling a particular space with sound can also be associated with a "shielding" effect at a metaphorical level. For example, the sound of the car radio has often been linked to protection—preventing the listening driver from falling asleep or developing a bad temper in heavy traffic. Advertisements visualize this effect by representing, for example, an umbrella with a musical score on its fabric, or musical angels accompanying the car—the umbrella and angels being metaphors for protection. At the same time, and adding another layer to our argument, car radio has also been welcomed for its masking effects, which turn sound production into a form of acoustical shielding.

Significantly, then, the rise of acoustic cocooning was anything but a straightforward process. The history of shielding interior automotive space from exterior noise and filling it with chosen sound, acoustically or metaphorically, had unequal and partially overlapping phases—not to speak of the differences in the sonic cultures of North America and Europe. And whereas we began and will end our story with the sound-image of the individual driver, alone in the car most of the time, we must not forget that in the early days of the automobile, drivers were rarely alone in the car. Though less consistently so today, at particular hours or on certain days of the week cars often carry more than one person, for example, parents and their children, or friends out for a good time.

This book reaches a number of conclusions we did not expect when we started our project. We knew from previous research that campaigns by noise abatement societies resulted in legislation by the mid-1930s, limiting the levels of noise emitted by car engines and horns (Bijsterveld 2008). We also knew that engineers strove to make engines quieter beginning in the 1920s, since noise came to be seen as a sign of mechanical problems and inefficiency (Bijsterveld 2006). But we connected interior car sound design with the 1990s. Indeed, that decade showed a spurt of interest in how cars sounded to drivers *and* to those who heard cars passing by—we have given some examples of this interest. Yet it is wrong to assume that interior sound design hardly existed before the late twentieth century,[7] for interior tranquility was on the agenda of automotive engineers as early as the 1920s. But why did these engineers bother with sound at all, and why at that particular moment in history?

MAKING SENSE OF SOUND

In 1929, the German engineering professor Wilhelm Hort published an article on street noise and options for its abatement in *Der Motorwagen*, an automotive trade journal (figure 1.2).[8] His article was a characteristic product of the 1920s. At that time, papers and pamphlets bemoaning the "nerve-racking" noise in big cities such as New York, London, Paris, and Berlin appeared frequently (Bijsterveld 2001). The most noteworthy aspect of Hort's paper, then, was not its topic, but its alphabetical list of ninety-one words that described a wide variety of sounds. The list started with words such as *ächzen, balzen, bellen, blöken, brausen, brodeln*, and *brüllen*, the onomatopoeia of which even

7. In an overview of the history of hearing, historian Jürgen Müller (2011: 18) also suggests widening our search for cases of sound design to the years before the most recent decades.

8. In 1914, Wilhelm Hort was chief engineer of the AEG turbine factory, as well as Privatdozent at the Polytechnic University (Technische Hochschule) of Berlin.

Figure 1.2
Cover of the German trade journal *Der Motorwagen* (1904)
Source: Der Motorwagen (1904), 7(1), cover.

non-German speakers can readily understand. It included sounds such as *donnern*, *dröhnen*, *gellen*, *girren*, *hämmern*, *heulen*, *klatschen*, *klingeln*, *knallen*, *krähen*, *lachen*, *meckern*, *pfeifen*, *prasseln*, *prusten*, *rollen*, *schellen*, *schlagen*, *schreien*, and *schmettern*. And it ended with *ticken*, *trampeln*, *trappeln*, *trommeln*, *tropfen*, *weinen*, *wiehern*, *wimmern*, *zirpen*, *zischen*, and *zwitschern*.[9] Hort indicated which sounds were directly related to street life: the barking of dogs, the trampling of horses' feet, the whiplashes of coachmen, the yells of street vendors, the clang of coal poured down a chute, the crashing of trash cans, the rumble of streetcars, the bang of automobiles, and the shrieks of their horns (Hort 1929).

Hort's list expresses the enormous variety in the sounds he heard, as well as his special interest in street noise. Today we do not instantly recognize all the sounds he identified. Nor do we automatically understand why they were a source of concern to him. Because some of the sounds—horses'

9. Providing translations that accurately express the meaning of these words is difficult, yet the following list gives an impression: groaning, creating mating cries, barking, bleating, roaring, bubbling, yelling, thundering, booming, reverberating, cooing, hammering, howling, clapping, ringing, banging, crowing, laughing, moaning, whistling, crackling, snorting, rolling, ringing, beating, crying, hurling, ticking, stamping, pattering, drumming, weeping, whinnying, whimpering, chirping (insect), hissing, chirping (bird).

trampling, the coachmen's whiplashes, and the coal's clanging—are no longer everyday phenomena of street life, his words for sound may not speak to us in the same way they did to early twentieth-century people. Car noise is still omnipresent, but new automobiles do not *bang* anymore. If concerns about vehicle noise survive to this day, the same does not hold true for street noise in all its various manifestations.

Our example points to the complexities of writing a history of something as ephemeral as sensory experiences, and about something as culture-bound as the meaning of sound. We do not have many recordings of the sound of early twentieth-century traffic, and when we have them—New York City commissioned such recordings in 1929—the quality of the recordings is so different from what we are used to that we may hear the recording equipment rather than the sound it was meant to capture. We thus usually have to rely on the texts contemporaries produced when discussing sound and noise. As we have pointed out in earlier publications (Bijsterveld 2008; Pinch and Bijsterveld 2012), this forces us to acquire a detailed understanding of the time- and context-bound vocabularies and conventions of speaking about sound. Experts, for instance, had different words for sound than laypeople had, as the automotive industry discovered when it tried to bridge the two vocabularies. To dramatize sounds as problems of noise, writers used conventional ways of describing sound—employing particular symbolic connotations of noise and silence, for instance—that displayed similarities across genres as different as novels, public hearings, and noise abatement pamphlets.

This book thus builds mainly on text and talk about sonic issues. First of all, we have interviewed over thirty experts involved in the automotive industry (from Ford, Opel, Renault, BMW, and Porsche), and in acoustic consultancy, sound design, noise control and standardization. In addition, trade journals like *Der Motorwagen*—the venue Wilhelm Hort published in—*Automotive Industries*, and the US-based *Journal of the Society of Automotive Engineers* allowed us to acquire a diachronic overview of the engineering perspective on car sound design, as well as advertising and marketing practices. Trade journals from domains such as the car radio industry complement this trade approach. Magazines representing the consumer's view on driving, such as those published and edited by intermediate consumer organizations and car manuals for the everyday driver, have been invaluable sources of historical information. To better understand the driver's perspective, we have also used, most notably for the early periods in this study, travelogues and literary sources.

Documents on legislative efforts at the national and international levels provided information on the regulations the automotive and civil engineering industry had to reckon with. Journals and archives of associations for car mechanics, archives of working groups within the International Standardization Organization (ISO) regulating the measurement of car noise, and of research groups focusing on the subjective experience of car sound enabled us to look behind the scenes of particular groups of experts who influenced the development of acoustic cocooning. Access to ISO files, which are not stored at ISO headquarters in Geneva but preserved by the working groups' secretaries and members scattered around the world, helped us to acquire a detailed and unique understanding of how the automotive industry responded to new societal demands concerning the sound of cars.

SENSORY STUDIES, MOBILITY CULTURE, AND HISTORY OF TECHNOLOGY

In interpreting the sources available to us, we have been greatly helped by numerous publications on the history, anthropology, and sociology of the senses. Within that domain, works on "auditory culture" or "sound studies" have been, of course, most relevant. Even more vast are the libraries on the culture of mobility and the history of automotive technology—similarly significant to our topic. Publications at the crossroads of these three domains were most inspiring and informative.

Which of these publications were most useful to our project and why? First of all, we would not even have recognized our topic as intriguing if we had not known the rapidly expanding number of works considering sensory orientations such as modalities of attention, thresholds of perception, and configurations of the tolerable and intolerable as historically and culturally variable. In fact, we owe much of how we phrased this issue to Alain Corbin and the French *Annales* school (Corbin 1995 [1991]). Corbin's study of "sound and meaning" in the nineteenth-century French countryside, for instance, shows that bells not only structured villagers' days and mediated news in ways we would not understand today, but also contributed to their spatial orientation and expressed the symbolic power of their towns (Corbin 1999 [1994]). His work on stench in eighteenth- and nineteenth-century France similarly illustrates how medical views, societal hierarchies, and cultural meanings associated with smell made French citizens use their noses completely differently than we do today (Corbin 1986 [1982]).

In the wake of such histories of everyday life and with a deep bow to the work of anthropologists, "sensory studies" has gradually acquired an important position within cultural scholarship. In such work the historicity and cultural conventionality of the senses—their categorization, order, hierarchy, meaning, and morality—is widely acknowledged (Howes 2005; Smith 2004; Pink 2009). Similarly important to our project is sensory studies' interest in "acoustemology," or epistemology grounded in sound and listening. This notion was introduced by the anthropologist Steven Feld, who studied the Kaluli in Papua New Guinea and their sensuous understanding of the world in terms of the sounds of their tropical rainforest (Feld 2003, 2005). In their "ear- and voice-centered sensorium," place and sense are interrelated: "as place is sensed, senses are placed; as place makes sense, sense make place" (Feld 2005: 179). Thomas Porcello has extended Feld's concept of acoustemology into "*techo*ustemology" to underline that technology also mediates our sensation, knowledge, and interpretation of acoustic environments (Porcello 2005: 270). Such work, in turn, has sharpened *our* sensivity to the ways in which the cultural appropriation of once new technologies such as the car body and the car radio rearranged the sensorium and acoustemology of the motorist (figure 1.3).

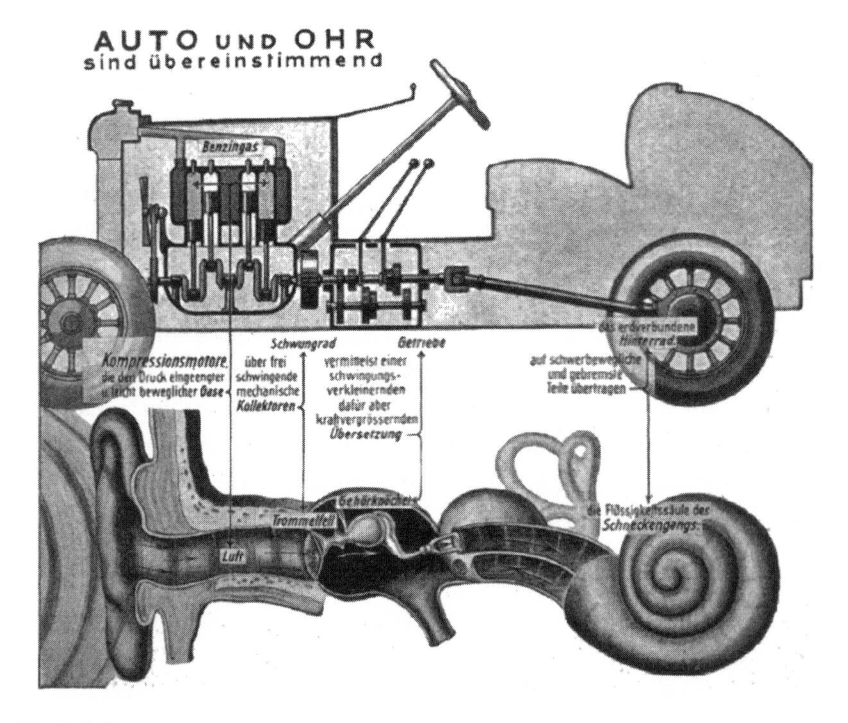

Figure 1.3
Popular science writer Fritz Kahn compared the form and function of the human ear with an automobile power train (1922–31).
Source: Debschitz and Debschitz 2010, 65.

Recent work on how new sound technologies make consumers experience their everyday lives in new ways (DeNora 2000; Bull 2000, 2007) has provided many examples of sensory enhancement and aesthetization. Such a take on the relationship between technology and sound, however, is rather different from an important strand of literature that started with Raymond Murray Schafer's work. In 1977, this Canadian composer and environmentalist set the stage for what we now call "sound studies" with *The Soundscape: Our Sonic Environment and the Tuning of the World*. In this seminal study, Schafer introduced notions such as "soundscape" and "schizophonia." The word "soundscape" referred to our sonic environment, conceptualized as both our everyday auditory environment and the musical compositions designed to improve it. The term "schizophonia" stood for "the split between an original sound" and its "reproduction" (Schafer 1994 [1977], 273).

Schafer argued that the Industrial Revolution, with its machines and intensified mobilization, negatively affected the "keynote" of the Western soundscape. Before industrialization became all-pervasive, Westerners lived in a hi-fi sonic environment: an environment with a "favorable signal-to-noise ratio," in which many single sounds could be heard clearly. The spread of the Industrial Revolution, however, has basically reduced this hi-fi environment to a "lo-fi" one, in "which signals are overcrowded, resulting in masking or lack of clarity" (Schafer 1994 [1977]: 272). This situation further deteriorated with the split between original sounds and their reproduction, itself the result of the introduction of the phonograph in the last quarter of the nineteenth century. The "aberrational effect" of these technologies contributed to the nervous *condition humaine* reflected in the notion of "schizophonia." To help the world re-create its sonic environment for the better, Schafer offered two solutions: ear cleaning, or training ourselves to listen more carefully, and a better acoustic design of our sonic environment. In order to advance this objective, he and his colleagues Hildegard Westerkamp and Barry Truax (1978, 1984, 1996) established and developed the World Soundscape Project in Vancouver. In their view, the new, improved sonic environment should be inhabited by sounds that please the ear.

Although audio technology often helped these scholars and activists to create sounding objects, Schafer's own attitude toward technology, as Ari Kelman (2010) has recently argued, remained highly ambivalent. Our approach shows more affinity with that of scholars like Jonathan Sterne (2003), whose approach to technology and sound is less normative. He has found inspiration in the work of the ethnologist Marcel Mauss and the sociologist Pierre Bourdieu, and has explored how with changing audile techniques—new connections between ears, bodily postures, technology, and practices—cultures of listening shifted in both enabling and constraining ways. For example, telegraph operators and early radio listeners had to learn listening through headphones as a focused activity *and* embed it in what was considered culturally acceptable and unacceptable behavior. We have similarly employed Mauss's and Bourdieu's insights to understand the changing audile techniques of car mechanics and their shifting attitude toward drivers' attempts to make sense of the sound of the automobile engine.

Even genuinely phenomenological work such as Don Idhe's book on listening and voice (1976)—with its stress on the spatial aspects of listening—and D. J. van Lennep's essay on the psychology of driving (1953) have been informative. We could only "listen" to Van Lennep by distancing ourselves from the ahistorical nature of his approach and acknowledging its roots in the 1950s. Once read in this way, his essay became telling. Van Lennep's analysis of driving as an imagined communication with fellow drivers from whom one at the same time keeps a respectful distance helped us to understand the changing experiences of motorists in a postwar world with increasing automobile traffic.

Of all work in the vast library on the history and sociology of transport and mobility, that of Wolfgang Schivelbusch (1979) on the railway journey, notably his analysis of how people struggle with their view of the landscape and the gaze and noise of fellow travelers, has been of particular relevance to our project. Also highly inspiring has been the study by Marc Desportes (2005) on the shifting experiences of traveling by coach, train, car, and airplane and the highway culture we live with today. Such work provided ideas for analyzing the historical relationships between the body, perception, mobility, landscapes, and culture. Concrete case studies with a similar focus, such as recent work by Thomas Zeller (Zeller 2010 [2006]), Mauch and Zeller (2008)), Cotten Seiler (2008), Jeremy Packer (2008), and Frank Schipper (2008) also helped us. Seiler, for one, contributed by clarifying the American origins of "our fantasies of the open road," while Packer has informed us about the culturally colored public concerns over traffic dangers such as road rage. The recent burst in the sociology and anthropology of mobility and movement, such as expressed in the seminal works by Mimi Sheller and John Urry—mentioned above—has been similarly informative.

To understand why our study cannot do without the cultural history of technology, let's return to the field of sensory studies. In 2000, Tim Ingold published *The Perception of the Environment*, in which he proposed to redirect research in sensory studies from "the collective sensory consciousness of society"—which had been the focus of David Howes (2005) and others—toward "the creative interweaving of *experience* in discourse" (Ingold 2000: 285, quoted in Pink 2009: 12, our italics). Sarah Pink, in turn, convincingly argued that it would be possible to undertake sensory ethnographies that would "both attend to and interpret the experiential, individual, idiosyncratic and contextual nature of... sensory practices, *and* also seek to comprehend the culturally specific... conventions... that inform how people understand their experiences" (Pink 2009: 15).

Since these claims may sound rather abstract, we will return to them at the end of the book with the help of the details from our stories. Here the quotations clarify why we need the cultural history of technology in particular—as a special branch within the wider area of science and technology studies (STS)—when it comes to translating Pink's principles in sensory ethnography into a *historical* approach to the car. Such work, after all, helps us to unravel the cultural conventions co-constituting the experience of technology. Kevin Borg's 2007 monograph *Auto Mechanics: Technology and Expertise in Twentieth-Century America*, for example, has pointed to early mechanics' skill in listening to the engine in order to detect flaws in its functioning, and why this sensory practice gradually disappeared: it was partially due to the US culture of organized distrust. Moreover, Leo Marx's *The Machine in the Garden* (2000 [1964]) has shown that literary writers in nineteenth-century America positioned the shriek of the locomotive, the rattle of railroad cars, and the thunder of engines as signs and symbols of dissonance—a deep disturbance of the pastoral ideal, a common cultural trope at that time.

Yet while Borg's and Marx's work has direct empirical relevance to our topic, other scholarship has been more relevant in terms of approach. For example, Rosalind Williams's *Notes on the Underground* (2008 [1990]) helped us perceive the cultural motifs structuring the reception of the roadside noise barrier, in which a nineteenth-century "underground aesthetics" seemed to return in the late twentieth century. Finally, the methodology used in *Driving Women: Fiction and Automobile Culture in Twentieth-Century America* (2007) by Deborah Clarke has been a source of inspiration. Her use of belletristic literature as a lens for studying automobile culture bears upon our second chapter, in which we analyze through the eyes of literati the cocooning experiences of motorists, among other issues.

In a recent essay, David Edgerton has complained about the fact that within the history of technology the "investigation of 'production' is shockingly démodé" (2010: 2). We will contribute to correcting this trend by studying both the consumption and production of car sound, with a special eye for the role of marketing departments as one of the intermediary "actors" in between. As we will show, marketers have reconceptualized the role of the senses and sound in selling cars in a way that has affected the research on car consumers as well as the process of car design.

HOW WE WILL NAVIGATE

Our navigation principles will be time and theme, as the chapters follow both a chronological logic and a thematic one. We will start in the 1920s and end in the 2010s, yet each chapter focuses on one particular phenomenon's contribution to the irregular rise of acoustic cocooning in cars. Although every chapter highlights certain decades, our thematic approach has occasionally inspired us to slightly widen such time-period scopes.

Chapter 2 starts with explaining the noisy character of cars in the 1920s and why this bothered only a few people in particular situations. It subsequently argues that enclosing car bodies and transforming the sporting automobile into a middle-class family vehicle—a process that occurred between the 1920s and 1940s—made automotive engineers, mechanics, and drivers listen to their cars with new ears. This resulted in the first initiatives to improve their interior sonic quality, whereby the marketing departments of car and car body manufacturers stressed that quietness was not only a comfort, but also a sign of the driver's societal standing. While this is an age-old trope in the cultural history of sound, another aspect of the early history of car sound design was fundamentally new. The introduction of the windshield and closed body not only introduced new sources of sound, such as creaking body parts, but also obscured the driver's view. Yet the first phase of acoustic engineering paradoxically saved the visual experience and tourist gaze in driving: acoustic smoothness compensated for the loss of the spectacular and, at the same time, reinstalled it.

Next, chapter 3 focuses on the 1920s to 1970s and unravels several processes of listening and de-listening to and in cars, which we group under the heading of "encapsulating the listening driver." This encapsulation had institutional and symbolic aspects, and once again changed the sensorial balance within the driving experience. Whereas handbooks initially taught drivers how to listen to the engine carefully in order to diagnose mechanical problems, listening to machines gradually became the sole jurisdiction of mechanics and automotive engineers, notably in Germany. Intriguingly, drivers' loss of their role in diagnostic listening coincided with the rise of the car radio. The context and meaning of listening to the car radio shifted dramatically over time, however. The way in which car radio was characterized changed from a safe companion on long rides in the country to a mood regulator that assisted drivers in remaining calm in heavy traffic, enabling them to withstand the impolite and rude behavior of fellow drivers. The ears of motorists, institutionally distanced from the sound of the engine because of their putative lack of expertise, became alternatively infused with the sounds of the car radio, like sonic drugs that helped drivers to cool down in the overheated traffic of the day.

Chapter 4 directs our attention to the noisy automobile on corridor highways. The chapter's main theme is sonic streamlining. During the 1970s and in following decades, the car and the corridor became subject to several forms of regulation focusing on sound. First, new national noise abatement legislation—encompassing several forms of noise instead of merely one source—and European initiatives to harmonize these laws increasingly informed exterior car sound design, notably noise emission, and the standardization of noise emission measurement. In addition, and again as a result of noise abatement regulation, more and more noise barriers began to be erected along highways. They soon proved to be a deeply contested highway technology—the interests of our ears competing with those of our eyes. As we will show, drivers' response to noise barriers was fueled by their appropriation of the roadside landscape as the private garden of their mobile home as well as by culturally shaped fears that resembled those connected to underground incarceration. The noise barriers elicited feelings of being locked in a closed space, of being trapped and buried. Together with an ever-increasing density of highway traffic, this fostered a situation in which audio technologies became increasingly important to the driver. Radio broadcasts with traffic and navigation information, played at a volume adapted to the traffic noise from outside the car, helped drivers to keep rolling in spite of a traffic jam or enabled them to assess options to escape it. Moreover, radio programs tailored to commuters and audiobooks "read" aloud made drivers more readily accept their fate if no escape was possible.

Chapter 5 concentrates on the rise of a new strand of sound design in the 1990s, aimed not only at creating silent interiors and low levels of noise emission, but also at making sounds that expressed both the make's and particular consumer group's identity and emotional needs. The search for these "target sounds" happened to be far more complex than the automotive industry initially expected because lay and expert listeners did not speak about sound in the same way. Yet the idea of providing each type of consumer a particular car sound, like the ringtone of a mobile phone, underlined the societal significance of an intimate and personally "tuned" interior acoustic space.

Our final chapter, chapter 6, returns to the questions articulated in the introduction. It aims to explain how the car became a place to relax with the help of sound and silence, and to provide insights that go beyond the psychological explanations suggested earlier. These answers draw on a cultural history of technology that incorporates ideas from the sociology of science and studies on the senses and mobility.

At two spots in the book we will temporally leave our main empirical focus on what we call the text and talk of sound. At the end of chapter 2, we insert a concise essay in illustrations that summarizes our main findings and shows their relevance for the decades after the 1940s. In chapter 5, we embed a section that can best be described as a YouTube essay on the commercialization of car sound, referring to period television commercials on automotive sound. We hope it will show our readers the significance of this trend *and* make them enjoy the often creative takes on car sound.

NEW THEMES ALONG THE ROAD

Scholars in sound studies have recently taken a step back from their beloved topic, stressing that "we" should not "hyposthasize" sound as something that needs our exclusive devotion at the expense of what we can see, smell, taste, and touch. (Personal communication Michael Bull, November 27, 2010). This claim echoes earlier recommendations in sensory studies not to either defend or attack the dominance of one sensory modality in particular societies or situations, but to start out from the "multisensoriality" of experience (Pink 2009: 19). This is, in fact, exactly what we noted when studying car sound and listening; we constantly stumbled on the intriguing relationships between hearing and the other senses, forcing us to think about the shifting historical relationships between the visual, auditory, and kinesthetic experiences of driving. We will claim, for instance, that the very earliest interest in car sound not only sustained a new rhetoric celebrating the ease, comfort, and distinction of driving, but also had to accommodate drivers' loss of view that came with the introduction of the windshield and the closed body. Early sound design intriguingly saved the predominance of the motorist's gaze by reducing the direct "feel" of the car through noise and vibration. Moreover, we will highlight the more recent parallel between the rapid spread of noise barriers, which have severely limited the visual dimension of driving, and the rising sales of audiobooks.

This does not imply, however, that we have not learned a lot about listening per se, or more precisely about listening as a cultural phenomenon. We have learned, for instance, to distinguish between modes of listening, one of which, for example, is diagnostic and monitory listening to the car engine. Observing the rise of the distinction between expert and lay listening prompted us to unravel how this distinction affected research on the subjective experience of sound and actual car sound design. Moreover, our project made us increasingly realize that listening to one particular "voice" usually came at the expense of listening to another—even though the connection may not be causal. When motorists started to invest less effort in listening to the engine, a new aural "attention span" for other sorts of listening emerged, yet it did so without determining *which* ones. As soon as listening to the car radio grew into an established activity, other forms of attentive listening lost their prominence or had to compete with the radio at particular moments. Listening tends to remain situational. Who listens to the morning news when driving a talkative colleague to a joint meeting?

This last remark, however, also opens up a new issue that we will return to at the end of the book. Even though we assume shifting modalities of listening, changing thresholds of perception, and time- and technology-bound aural techniques, we also seem to evoke a *timeless* theory of the necessity of sensory balance. How can we *theoretically* account for such a coexistence of assumptions and approaches? Is it possible, for instance, to connect perception psychology and sensory history in interesting ways, without doing injustice to either one of these domains of inquiry?

Throughout the project, existing roads did not lead us to all destinations; we had to build highways and install traffic signs when traveling across unknown terrains. We constantly had to invoke two particular strands of theories as "signs" to understand what happened to the sound of cars. These were, first, theories on the increasing personalization, "aestheticization," and "emotionalization" of commodities in today's experience society and its style of emotional capitalism—for which we relied on work by Gerhard Schulze (2005 [1992]), David Howes (2005), and Eva Illouz (2007, 2009)—and, second, theories on harmonization and standardization. At first sight, the phenomena of personalization and standardization appear to contradict each other: standardization seems to make things more impersonal and void of particular emotions. We will show, however, that both these phenomena have affected the rise of acoustic cocooning. Standardized production enabled personalized combinations of uniform units, varying from modular systems in noise barriers that had to express the *couleur locale* of a particular place, to sets of recombined digital sounds expressing the identity of particular groups of consumers. But standardization also threatened to reduce the options for embedding sensory stimuli in car design in such a way that existing sensory designs were kept from extinction and new sensory designs became sought after, as our behind-the-scenes look at international standardization will show.

In the conclusion of the book, we consider what has fueled us from the beginning: a common interest in technology and culture, or, more specifically, in how members of particular societies produce new sound-related technologies and culturally appropriate them into the practices they want to sustain. As Mikael Hård and Andrew Jamison have explained, such processes of cultural appropriation are about "using and learning and interpreting and giving meaning" to new technologies (2005: 4).[10] This observation prompts us to reflect on the implications of these practices. What have been the consequences of acoustic cocooning for what we consider private and public space on the road, to mention just one issue? And what does such a shell imply for our notions of responsible driving? What does, in other words, driving "sound and safe" mean for arriving "safe and sound"?

10. For cultural approaches to the history of technology see also Pursell (2010).

CHAPTER 2

"It Shuts with a Comforting Sound"

Closing the Car Body

A SOLID SOUND

In 1955, General Motors (GM) issued a short industrial film about the car bodies Fisher produced for the car manufacturer (figure 2.1).[1] *Body Bountiful* was the film's title, clearly meant to attract a wide audience. The point GM wanted to convey was that the mass production of car bodies is an upgraded version of the tradition of craftsmanship in manufacturing. The movie highlights this craftsmanship through the many details and skills involved in the production of the car body. One such detail is the sound of the door, demonstrated by a woman who opens and closes it several times, with a sonorous male voice-over saying: "It shuts with a comforting, solid sound." The film goes on to explain the process of door design and how the Fisher labs perform extensive tests on the opening and shutting of the doors.

This scene reveals that sound is hardly a new concern in the automotive industry. Such attention to sound has actually been part and parcel of car design and marketing since the 1920s,[2] although it took on a

1. "Body Bountiful" (1955), 25 minutes, available at http://www.youtube.com/watch?v=NbquyC8Xz5Y (accessed September 3, 2012). The car door scene can be found between 04:50 and 06:42 minutes after the start of the movie. Please note that the official name of the Fisher company was "Body by Fisher." In this chapter, however, we will often use the shorthand "Fisher," as the company itself often did as well.

2. The earliest advertisement referring to "silence" we have found so far dates back to 1912. It is an ad for a 1913 Peerless Seven-Passenger Open Touring Car, manufactured by the Peerless Motor Car Company, Cleveland, Ohio (United States), and was published in the *Review of Reviews Motor Department*, 1912, 73.

Figure 2.1
Still from the industry film *Body Bountiful* (1955)
Source: http://www.youtube.com/watch?v=NbquyC8Xz5Y, accessed September 3, 2012.
Courtesy: General Motors LLC.

new significance and twist in the 1990s. Since that decade sound design has focused on noise control as well as on constructing brand-specific sounds and "target sounds" for particular groups of consumers, but early efforts in automotive sound merely concerned the reduction of noise and vibration.

The film's solid-sound scene also suggests that bodies affect car sound in important ways. This chapter explains the historical origins of this connection. It shows that in the car's transformation from a motorized open coach to a sealed box on wheels, automotive engineers, manufacturers, and drivers started to listen to the sounds of cars in new ways. In this context, the years between 1920 and 1940 were a crucial period. We will explain how and why the car-manufacturing industry promoted mechanical, convenient, and aristocratic silence as distinct conceptions of sound and what this orientation meant within contemporary car culture. The pursuit of silence reflected an ideal more than a reality, but, as we will argue, the silencing effort sustained the highly valued *visual* experience of the driver. By reducing the levels of vibration and noise, this effort preserved the tourist gaze associated with early automobility—a valued feature that came under threat after the introduction of closed bodies.

Our argument will draw on the automotive trade and engineering literature from the United States, France, and Germany, as well as the rich historiography about the interwar years in automotive history. Car owners in the United States were the first to prefer the closed "sedan" to the open tourer (or "phaeton") on a large scale. They were soon followed by motorists in the United Kingdom, where the sedan was known as the "saloon." The situation in France is especially noteworthy because one of its manufacturers introduced a unique approach to quieting the car body. This had a lasting effect on the industry, not so much through imitation of the design itself, but by contributing to the debate about interior car sound. We will therefore elaborate on French car body manufacturing in particular.[3] The heydays of the German automotive industry had not yet arrived (Binnebesel 2007), but the German examples of manufacturing trends, especially in the later sections of this chapter, serve as enriching extensions of the French and American ones. In addition, we analyze travel accounts, travelogues, and literary representations of motoring in order to open up the auditory and other sensory experiences of driving. Such sources help us understand what it meant for drivers and passengers to be enfolded by walls and windows when they had been accustomed to feel the wind and weather and hear the engine roar.

THE RISE OF THE CLOSED SEDAN: SOUND BODY BUILDING

At the end of the nineteenth century, motorists could indeed hear the car engine roar. It was quite noisy machinery. Because the internal combustion engine was based on the controlled explosion of a mixture of air and gasoline compressed in a cylinder, the early gasoline engine banged tremendously so long as combustion remained uneven. This meant that the gasoline car made more noise than the cars running on steam and electricity (Volti 2004: 7). However, for the men who drove a gasoline car then—wealthy engineers, physicians, and businessmen—its noise was hardly a problem. On the contrary, noise stood for the power of machines, and allowed drivers to impress bystanders (McShane 1994: 169, Mom 1997: 475). In addition, noise informed motorists about the condition of

3. The body of literature on French (auto)mobility is very rich, yet the history of the French car body remains underresearched. Jean-Henri Labourdette, grandson of coachbuilder Henri Labourdette, published on the French car body (Labourdette 1972), yet his study does not meet contemporary quality standards for the history of technology. For the historiography of French automobility, see Dumont 1973; Bardou et al. 1982; Loubet 1990; Laux 1992; Loubet 2001; and Merki 2002.

the car: uncommon or loud sounds articulated an urgent demand to have it maintained or repaired.[4]

Only in specific contexts did the noise of the gasoline engine prove to be a concern. In the United States, for instance, it was not deemed acceptable to use an automobile in funeral ceremonies before the 1930s (Berger 1979: 183). It was not that bystanders did not notice and deplore the car's loudness. In *The Life of the Automobile*, a famous 1929 critique on capitalism and the speed of modern society by the Russian journalist and writer Ilya Ehrenburg, the arrival of the car in Paris is full of noise:

> With a disrespectful roar it leaped out into the sleepy boulevards. Old nags reared and pranced, and terrified ladies pulled vials of smelling-salts from their *réticules*. The automobile moved along in fits and starts. It hopped like a kangaroo. It would jam and then unexpectedly jerk forward again. The streets were filled with a horrible stench. The din was louder than a thunderstorm in spring. (Ehrenburg 1999 [1929]: 14)

But car noise was not yet widely considered a public problem in need of reduction. Only during the second half of the 1930s did protests against the noise of urban traffic reach such an intensity that reducing the car's exterior noise became a goal in its own right. As a consequence, in large parts of Europe, it became mandatory to equip the exhaust pipe with a muffler and to limit the sound levels automobiles, including their horns, were allowed to make (Bijsterveld 2008).

These interventions focused on what urban residents and people on the streets could hear. However, car manufacturers had a decade earlier already begun to tackle the sounds that the *driver* was confronted with. The main reason for their concern was the quick shift from open to closed cars. This new emphasis on sound may seem somewhat surprising. Didn't the closed body shield the driver and other inhabitants from the noise of the wind? And didn't it block out engine noise, at least partially? The answer to both of these questions is yes. But the closed car created new noise problems. What's more, the transition from the open phaeton to the closed sedan expressed a shift in patterns of use of automobiles—a change that manufacturers both sought and responded to by establishing a new sonic ideal for the car: that of "silence." Discourse about the sound of car bodies was strikingly similar across the United States, Germany, and France. In France in particular, however, articles in engineering journals, ads by chassis and

4. For more on this in Germany, see our next chapter. For French accounts, see Anonymous 1920, 1924a.

body manufacturers, editorials, driving tests and readers' letters in consumer magazines, and customer surveys all advanced the message that the best car was a "silent" car (Krebs 2011; Bijsterveld and Krebs 2013).

Let's focus on the closed car revolution for a moment to unravel both the rise of new sonic standards for the automobile and its particular significance in France. The shift from the open to the closed car was swift and decisive (Rolt 1950, 1964; Nieuwenhuis and Wells 2007). While in 1919 only 10 percent of the cars built in the United States had a closed body, the production numbers for closed cars surpassed those for open tourers in 1925. Two years later, over 80 percent of the cars manufactured in the United States were of the closed type (Flink 1988: 213–14). With some delay, a similar development occurred in Europe: from 1926 onward, the closed sedan became the preferred model in the United Kingdom, France, and Germany. By the end of the decade, it dominated these car markets (Rolt 1950: 113; Edelmann 1989: 98).

Contemporary commentators and historians of mobility have interpreted this dramatic change in technology as an expression of shifting patterns of use.[5] This shift is best captured as a process of civilizing the automobile, converting it from adventure machine to family commodity, a development we briefly flagged in chapter 1. Prior to World War I, the automobile had been a plaything for wealthy upper-class sportsmen: touring and racing were the chief pursuits, and touring cars or runabouts with folding tops were the dominant body designs. In the interwar years, however, the car became a basic means of transportation for members of the upper middle class. To accommodate both business trips and family outings, the automobile needed to be reliable and easy to handle. Additionally, it had to provide comfort and weather protection in all seasons—the basic features of the permanently enclosed owner-driven sedan (Lambert 1927; Mom, Schot, and Staal 2008). Its solid top was meant to prevent "wind, dust, and rain" from blowing "into the passenger compartment," and ruining one's coiffure (White 1991: 58). "If you really want your wife to drive," ads from the 1920s advised readers, buy "year-round comfort" for the family (advertisement, Chandler 1924; advertisement, Jewett 1924). The closed cabin meant a reduction in the number of passengers, from seven or eight to four or five, and "united" drivers and passengers, as the German journalist Robert Allmers noted, "in one interior" (Allmers quoted in Edelmann 1989: 94).

5. For contemporary comments see Petit 1922a, and R. B. 1924. For historical accounts see Flink 1992; Koshar 2002; and Mom, Schot, and Staal 2008.

Initially, car manufacturers built only the engine and chassis, while specialist coachbuilders made the body, a wooden frame covered with wooden panels. The earliest body-engineering techniques "were literally carried forward from the days of the horse-drawn carriage" (Ludvigsen 1995: 52). Shortly after World War I, however, engineers and journalists began to criticize the traditional construction methods as old fashioned, impractical, or even "true anachronisms" (Petit 1922b: 155; Anonymous 1921). Painting the wooden car body, for instance, took weeks because it required applying "coats of lead-color, filler, stopper, stain, ground color, and body color, covered and protected by flatting and finishing varnishes" (Ludvigsen 1995: 53). Two alternative design principles competed with the traditional wooden body: a composite construction introduced before the war, which had a wooden frame covered with steel panels, and the all-steel body.

In 1921, the French aviation pioneer Charles Weymann added a second composite design. Adopting methods from airplane construction, he padded the wooden frames with synthetic leather. The body had an ash frame held together by steel plates instead of the mortise-and-tenon joints used by other coachbuilders. These techniques allowed Weymann to build "supple, workable, light and silent" bodies (Anonymous 1925a: 539). A closed sedan for four passengers weighted only 200 kilograms, half the weight of an ordinary open body (or "torpedo") of that time (Petit 1922b: 157). To advertise his design, Weymann equipped a number of high-end European makes such as Voisin, Panhard, Hotchkiss, and Delage with his bodies. As a Hotchkiss ad shows, it was portrayed as "souple" and "silencieuse" under all conditions (figure 2.2). In addition, Weymann sold production licenses to other manufacturers and coachbuilders. By the mid-1920s, he had about forty French licensees, one of which was Renault, and about the same number in other European countries (advertisement, Weymann 1926).

Eight years after the introduction of his airplane-inspired wood-and-leather body design, Weymann again came up with something novel: the semirigid body. This composite body did not have fabric between the wooden frames, but steel plates. This innovation sparked a heated debate about effective body building in the trade journal *La Vie Automobile*. Proponents of the flexible body, one of whom was the prominent automotive journalist Henri Petit, pointed out that it was far more quiet than the semirigid one (Petit 1929). Yet the semirigid body was thought to be the safest because it provided better protection in accidents (Charles-Faroux 1929a). At the other side of the Atlantic, the businessmen behind Fisher Body believed there was no need to choose between safety and silence. They were convinced that "wooden frames sheathed in steel" brought "superior strength" *and* "quietness" (White 1991: 63). Weymann's opinion on the issue ceased to matter by 1930, however,

Figure 2.2
Hotchkiss Advertisement (1928)
Source: Advertisement, Hotchkiss 1928.

when he ran into financial trouble during the economic crisis and was forced to liquidate his firm.

In the meantime, Weymann's French competitor André Citroën had acquired the American Budd Company's license to make all-steel bodies. The all-steel design promised easier mass production and enhanced styling options, and also enabled Citroën to advertise himself as the first French car manufacturer to use modern American production methods like the assembly line and interchangeable parts (Laux 1992: 77). In 1927, a Citroën advertisement proudly announced that 50 percent of the cars it produced daily had all-steel bodies. The ad not only promoted the all-steel body as safe—"the resistance of its steel panels is a guarantee against the dangers of the road"—but also as silent, probably in response to Weymann's earlier claims about the silence of his flexible bodies (advertisement, Citroën 1927).

In the course of the 1920s, the *trois grands* car manufacturers, Citroën, Renault, and Peugeot, all developed their own body production capacities. Body production thus increasingly became an activity of the car manufacturers themselves. This and the economic depression severely diminished the number of independent coachbuilders in the 1930s (Principaux Fournisseurs 1928, 1936; Bardou et al. 1982: 141). Even more decisive was Citroën's introduction in 1932 of the *monopièce*, a technology that integrated chassis and body in a single unit, effectively ending the traditional distinction between car manufacturers and coachbuilders (Loubet 2001: 105). From this point onward, automotive engineers had to deal with chassis and body as one unit (Maillard 1935b)—a thing that fully enclosed the driver.

MECHANICAL SILENCE AS A SIGN OF RELIABILITY AND ENGINEERING QUALITY

We have just seen that the issue of the "silent" driving experience acquired prominence in the automotive world with the rise of the closed sedan. Yet we haven't explained *why* the issue of sound became so important in the context of new car design.

We have found one of the answers in the need to make automobiles easier to handle and more reliable in order to reach middle-class customers—who were usually less well educated in car technology than the motoring pioneers. An important way of attracting them was by delegating some of the driver's tasks to the machine itself. This redefinition of the driver's skills implied, in the words of cultural historian Kurt Möser, a process of "de-sportification" (*Entsportlichung*) (Möser 2009). It started with replacing the levers of the ignition at the steering wheel by pneumatically assisted adjusting mechanisms beyond the driver's control. Another step was the synchronization of the gearbox, which rendered superfluous tedious double clutching when gearing down. Automation also manifested itself in the replacement of the oil pressure gauge by a simple warning lamp, inviting the motorist to act as soon as it lit without knowing the reasons for the emergency, and in automatizing the lubricating and timing of the engine, which had previously been handled manually (Petit 1921).

Women in particular were expected to appreciate easy-to-handle cars (advertisement, Berliet 1926; advertisement, Repusseau 1926; Scharff 1991). Thus, in the interwar years, the driver's tasks shifted from controlling the machine to driving it (Möser 2004: 96, 2009: 164, 178; Mom 2007). Some drivers mourned this assault on their technology-related automotive competencies, but it ended in the engineers' victory with the "electronics revolution" in the 1970s (Mom 1995; Kehrt 2006: 215).

Delegating drivers' tasks and skills to the machinery and making the car less prone to deficiencies wasn't enough, however. Drivers had to learn to *trust* their automobiles, and sound had an important role to play in this sense of ease, since the right sounds could enhance the *auditory* impression of *reliability*. The earliest articles on noise in trade journals warned that "knocking" engines and similar sounds worried drivers and made the less experienced believe the engine had a major problem even when it didn't. Noises such as these induced "anxious feelings" (Anonymous 1904: 58; Anonymous 1905). As Ilya Ehrenburg noted in his documentary novel on the life of cars, the car buyer of the interwar years was "high-strung" and demanded "a noiseless engine" (Ehrenburg 1999 [1929]: 27). Quieting the car, Henri Petit contended, avoided noises that might lure an uneasy driver into wrong diagnoses of engine problems (Petit 1935). We will return to the phenomenon of listening to the engine in our next chapter. At this particular point in the story, it is important to understand that engineers especially wanted to get rid of sounds that could mislead the driver. One way of doing this was to silence the mechanical components of the transmission, notably the mysterious howling and crunching of gears, or, as one engineer called it, "the clutch and transmission jazz" of a car (Vincent and Griswold 1929: 388).

Still, the industry's early interest in noise control did not stem exclusively from the aim of reassuring new cohorts of customers. It gratified the ears and esteem of the automotive engineers themselves as well. As historian Allard Dembe has explained, at the turn of the century, engineers in general began to reduce the noise created by equipment after realizing that "noisy machinery often is an indication of mechanical inefficiency that ultimately can result in lower productivity and increased cost" (1996: 195). Mechanical noise thus became a sign of wear and tear and a *loss* of power, at least for engineers. Such ideas were taken up by the automotive industry even before the heyday of the closed sedan. The German trade journal *Der Motorwagen*, for instance, ran several series of articles on sound-related issues during the first two decades of the twentieth century, and time and again, the message conveyed was that noise signaled a war of attrition, although knocking engines also involved chemical complexities.[6] In 1920, the American Hoover Steel Ball Company explained how it had researched "the problems of friction" in car engines, and how their steel balls provided the answer:

> Inside of the smoothly gliding mechanism of the motor car is the well-poised ball bearing insuring locomotion luxury that is free from sounds which offend the hearing and shorten the life of the unit. (advertisement, Hoover 1920: 61)

René Charles-Faroux, editor in chief of *La Vie Automobile*, went so far as to define silence as the symbol of the car's mechanical quality. "Silence," he declared, "is for sure among the finest qualities of an automobile. Mechanical noise always represents a loss of energy" (1933: 169). Mechanical silence had thus begun to represent *engineering quality*.

As we have already mentioned, gears were suspect number one: they had a reputation as "the noisiest" components of the car (Petit 1928: 325). An American consulting engineer analyzed the four "characteristic sounds" of gears: intermittent clicking, an irregular growl, a pulsating growl—known as "run-out sound"—and a high-pitched squeal (Buckingham 1925a: 62). He argued for a novel composition of "harmonious sounds" and for treating the transmission as a "violin body," while another engineer announced that the time had come "to design a car with a tuning fork" (Buckingham 1925b: 461; Horning 1925: 93). Even if engineers did not become first-class composers, the slapping sounds and swishing of oil coming from the transmission were tackled by reducing tolerances in production, improving sound absorption (for instance, by using leather and felt), and changing the sound itself (Snook 1925; Firestone 1926; Herrmann 1922; Stanley 1926). One frequently discussed intervention was the replacement of gears by chains to connect the camshaft and the crankshaft (Petit 1928). "Morse, of course," was one of the industries equipping passenger cars with "genuine silent chains" (advertisement, Morse 1932: 64). But the gears themselves were also altered to reduce noisiness, with flexible materials like rubber or cotton-reinforced Bakelite, sold under brand names such as Textolite and Celoron. "The conquest for silence moves on" read the headline of a Celoron ad for "silent timing gears" (advertisement, Celoron 1932).

6. These series discussed the phenomenon of "knocking" engines: Anonymous 1904; Anonymous 1905; Richter 1925, 1926a, 1926b, 1926c; Heller 1926; and Wedemeyer 1929; the rise of silent gear chains: Praetorius 1913a, 1913b, 1913c; Anonymous 1916; and issues such as the noise of cogwheels and the use of spiral springs: Wetzel 1914, and Wilkens 1918a, 1918b.

Balancing the rotating structures of the internal combustion engine was another way to address noise. Six- and eight-cylinder engines functioned better than four-cylinder engines in this respect, for the simple reason that bigger engines "combine a good torque at low speed with smoothness, silence and flexibility" (Rolt 1950: 97). Such larger engines also allowed drivers to stay in direct drive most of the time—the only genuinely silent gear. Moreover, to "reduce the necessity of gear shifting, engines with a volume of three to four liters are preferred" (Charles-Faroux 1929b: 577).

There was one big disadvantage of the oversized engines, however: the half-closed throttles in the inlet channel increased fuel consumption. American engineers, living in a country with cheap gasoline, were ready to pay that price and even developed a costly automatic transmission when engine speed started to rise and urban driving threatened to bring back frequent gear-shifting (Gott 1991; Mom 2008b). In contrast, the French automobile taxation regime favored small, high-revving engines because cars with big engines were viewed as luxury objects that should be heavily taxed (Bardou et al. 1982). The French automotive industry therefore had good reasons to search for ways to control engine vibrations beyond extending the number of cylinders. In 1931, Chrysler introduced a promising solution: "floating point" technology. This two-point engine mounting system used flexible rubber mounts to prevent vibrations from reaching the frame and body. A four-cylinder engine with this mounting system, Chrysler claimed, offered the "smoothness of an eight with the economy of a four" (Weiss 2003: 145). Soon after, Citroën invested $250,000 to acquire a license from Chrysler and began equipping the new Citroën 6G with a *moteur flottant* (Loubet 2001: 105).

Making the car *body* mechanically silent proved to be quite another issue, however. Initially, cars with closed bodies were not perceived as quiet. On the contrary, in 1925, an anonymous author complained that the closed body often caused such objectionable noises that an open car was preferable (Anonymous 1925b). One explanation for these problems is that the body made sounds audible that had hardly been noticed when in an open tourer. In the open car, air resistance, wind, and traffic masked sounds that suddenly betrayed their existence the moment the closed body raised a barrier between the motorist and the outside environment. Even today, the sounds hidden by masking challenge engineers and move them to pursue sonic perfection: as soon as they have silenced one source of sound, they hear and try to abate others (see chapter 5). Another explanation for novel sonic concerns with the closed cabin is that the enclosed space sharpened the auditory focus of the driver, in part because of the diminished view. We will come back to the relationship between the auditory and the visual in the penultimate section of this chapter. For the moment, it is sufficient to understand that silencing the space within the closed car body required a special effort.

The interventions made depended on the type of construction. Composite bodies, for instance, had very specific squeaks and rattles. The French coachbuilder Audineau tried to eliminate them by applying rubber and artificial leather bumpers between the body and chassis and between the wooden frame and metal panels. To convince consumers that the closed car moved more smoothly through space "than a torpedo," Audineau ads used engineering drawings to point out the numerous spots where special measures had been taken to silence the body (advertisement, Audineau 1925a, 1925b; Charles-Faroux 1931) (figure 2.3). As we have seen, the Weymann fabric body was also promoted as being silent. The engineering secrets behind its silence were the flexibility and lightness of its construction. In the 1960s, the British mechanical engineer Tom Rolt remembered that "the genuine Weymann body was handsome, durable and relatively rattle-free," adding that its "cheap imitators proved quite the reverse and

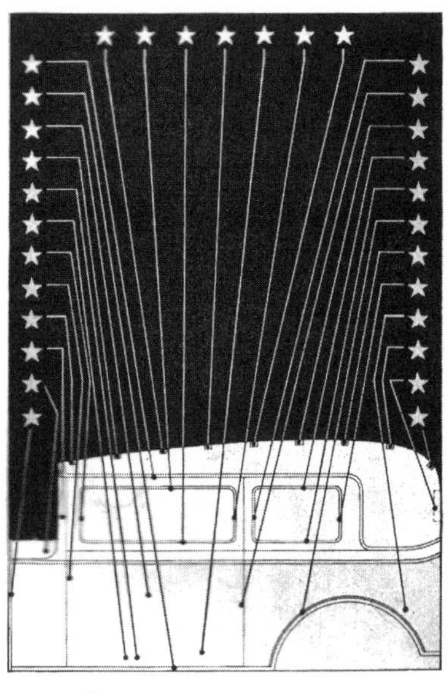

Figure 2.3
This detail from a 1932 Western Felt Works ad indicates the thirty-one places where felt could be used to reduce noise
Source: Advertisement, Felt 1932.

brought the fabric body into disrepute" (1964: 115). In making this claim, he not only testified to the typical Weymann flexible-body sound, but also provided an explanation for the downfall of the fabric body: it was abandoned for safety reasons *and* as a consequence of the negative publicity that followed from the low-quality imitations.

Contemporary commentators acknowledged that the semirigid composite bodies that succeeded the fabric body did not match the original Weymann silence: "We have to say that almost all bodies are noisy upon sale" (Charles-Faroux 1933). In their view, the steel and wood-steel bodies could not be silent *par construction* (Maillard 1933). It is not surprising that manufacturers producing and using wood-steel bodies such as Fisher and Renault fiercely disagreed (advertisement, Renault 1931). And so did Citroën, which licensed Budd's all-steel body manufacturing process, citing "the absence of joints" as key to the closed all-steel body's silence (figure 2.4) (advertisement, Citroën 1927). Citroën did not broadcast, however, that the engineers of steel bodies had to cope with new noise issues in exchange for their greater strength. The large steel panels transformed the body into a sort of boom box that amplified the vibrations from the chassis and drive train. One engineer claimed that the steel body's sound was better captured by the adjective "resounding" (*sonore*) than the word "noisy" (*bruyante*). Applying special varnishes helped dampen this amplifying effect, but in this respect the French manufacturers were way behind their American competitors (Maillard 1937).

Moreover, as already discussed, any progress in silencing the car body was quickly made redundant by successes in engine noise control that reduced the benign masking effects of noise (Charles-Faroux 1929a, 1929b). In turn, quieting the body revealed engine noise and other sources of unwanted sound. Charles Brull, former head of the Citroën research laboratories, had an original take on this process: "It has not been for very long that the free concert of the body successfully masked the noise caused by a poorly balanced, high-revving engine. Cynical colleagues say that for the same reason they cannot wait for the widespread use of car radios" (Brull 1935: 259). A more fundamental solution than turning on the radio was the ubiquitous application of flexible materials like the ones we already mentioned when discussing gears (Wolff 1935). In addition, rubber helped to avoid metal-on-metal contact. In France, the company Silentbloc offered rubber applications for a variety of car components, including engine mountings, springs, brakes, gearboxes, and body frames. The company's 1926 advertising campaign announced that widespread use of their applications would

Figure 2.4
Budd advertisement (1932)
Source: Advertisement, Budd 1932.

turn the next automobile fair into a "Salon of Silence" (advertisement, Silentbloc 1926).

Quietness was never really silence, though, and research to achieve it intensified in the 1930s (Kagan 1937). No doubt this was connected to the rise of noise abatement societies, campaigns, and legislation, as well as to the introduction of units such as the phon and the decibel for measuring loudness and sound intensity in the mid-1920s (Bijsterveld 2008). Engineers started applying new tools, such as acoustic test stands (Charles-Faroux 1937), and the introduction of "objective" acoustic measuring devices, substituting for the engineer's ear, was seen as a crucial step toward noise control. In 1932, the French engineer Marc Chauvierre stressed that "if you want to study a certain phenomenon, you have to be able to measure it in the first place" (Chauvierre 1932: 1650). In subsequent years, speakers at the conferences of the French Society of Automotive Engineers (Société des Ingénieurs de L'Automobile, SIA), continued to present the latest electroacoustical knowledge and acoustical instruments in automotive research (Brull 1935; Kagan 1937).

We will reveal the results of such studies below. Here we want to conclude that in the interwar years, mechanical silence had become a new landmark of engineering excellence. This is perhaps best illustrated by the rise of a new benchmark in the test drive reports published in *La Vie Automobile*. The authors of these reports began to specify their "auditory experiences" as a novel category for judging and expressing the engineering quality of the automobile chassis and body (Petit 1923a, 1924b; Chauvierre 1926).

CONVENIENT SILENCE AS IDEAL FOR LONG-DISTANCE TRAVELING

The increasing reliability of automobile technology enabled motorists to make long-distance journeys without having to worry about engine trouble or wheels falling off en route. First, they used the car to drive around town, then to make weekend trips, before graduating to long-distance drives (Kline 2000; Interrante 1983; Belasco 1979; Bertho Lavenir 1999). Car tourism contributed to the articulation of a new "companionate family ideal," in which the father, typically the sole driver on family outings, sought to restore Victorian values of "mutual, voluntarily given affections." To this end, the "canned family" celebrated the car as being "more homelike than home itself" (Belasco 1979: 55, 57, 67, 68).

The rise of long-distance driving in a mobile home away from home triggered a call for a new type of comfortable silence we will call "convenient silence." The nonstop noise that drivers were confronted with during prolonged travel was considered exhausting. One automotive journalist with a talent for drama declared: "After a journey of five or six hours in a noisy car you are literally anaesthetized, your ears are buzzing, [and] your thinking is destroyed" (Maillard 1933). While tourism in the early age of the automobile had been about making short trips to enjoy the beauty of the scenery when the weather was good, the new style of tourism was more about getting directly from place A to place B. For touring of this sort, the closed sedan was preferred to the "torpedo" because, as Henri Petit explained, "the closed body protects against the familiar noise produced by the whistling of the air, notably at full speed," and "muffles . . . most noise of the chassis" (Petit 1924a: 197).

The kind of body quietness Henri Petit was referring to differed from the mechanical silence discussed in the previous section. This was not the quietness of a body that abstained from creaking or squeaking, a body that *kept silent*, so to speak. Petit was referring to the silence created by a space that insulated its inhabitants from outside noise, a body that *protected against noise*. This silence compensated for the lack of the chassis's mechanical silence. We can also deduce this from the praise Petit gave to the Weymann body: it offered comfort to the passengers. "This comfort results from its tight silence. And silence is the quality we are longing for, especially in an automobile, even more when the automobile is used for long-distance travels. . . . Because the occupants are completely insulated from the chassis by the floor panel and three layers of fabric, one of which is a thick layer of felt, they can hardly hear any exterior noise." This insulation or shielding from exterior noise also helped drivers and passengers to stay fresh, since "the fatigue of passengers after a long journey in a closed car results from the constant noise that penetrates their ears. In a Weymann this noise is perfectly muffled" (Petit 1929: 239).

Just like mechanical silence, the absence of drivers' fatigue after a long trip became an important feature for judging the quality of a car. In 1923, a French test driver was offered the unique opportunity of riding in a car fitted with two kinds of body: a torpedo and a closed body. In his evaluation report, he explained that while the chassis was identical in both cars, the one with the closed body had a significant advantage, thanks to its convenient silence, of not causing any fatigue (Lefèvre 1923). Another test driver expressed his experiences in a closed sedan in the following way: "When I arrived, I was as fresh as I had been at the start, without the slightest buzzing in my ears, as if I stepped out of a well-cushioned living room, instead of an automobile" (Anonymous 1924b). The metaphor of the closed car as a living room on wheels referred to other aspects of automobile comfort as well, such as the soft upholstery, fine fabrics, accessibility of the rear seats, and the protection the body provided from bad smells and mud (advertisement, Chrysler 1927; Charles-Faroux 1935; Maillard 1935a). But sound had its own distinctive role in the construction of car comfort.

Figure 2.5
Labourdette advertisement (1925)
Source: Advertisement, Labourdette 1925.

In 1922, *La Vie Automobile* asked its readers[7] to rank ten automobile qualities according to how much they influenced their choice of car. Most of the 2,247 respondents put either "endurance" or "economy of operation" highest on their list. "Comfort" was third, followed by "hill-climbing capability," "price," and "speed" (Charles-Faroux 1922). French manufacturers took the results seriously and began highlighting comfort and convenient silence in their marketing campaigns. For example, Henry Labourdette's 1925–26 *Silensouple* campaign featured two men in a car, one in a relaxed pose with his finger touching his ear as if listening attentively, and the other stretched back in the seat with a cigarette in one hand, as if re-creating at home (figure 2.5) (advertisement, Labourdette 1925). This was acoustic

7. *La Vie Automobile* was a leading, weekly automobile journal for a broad middle- and upper-class audience of automobile enthusiasts in France. René Charles-Faroux was its editor in chief and the doyen of French automobile journalists.

cocooning for the higher-class owner. There were also ads promising convenient silence for low- and medium-priced cars, however. A 1926 Renault ad promoted its 6 CV and 10 CV models for "true automobile comfort." These models' "mass-produced body dampens all vibrations without noise or resonance, and its silence eliminates fatigue" (advertisement, Renault 1926a). Voisin, a competing car manufacturer, stressed the importance of silence in yet another way. It reminded its prospective customers that "of all sensations the human organism has to endure, noise is the most oppressive" (advertisement, Voisin 1927).

The deep interest of the car-manufacturing industry in the sonic qualities of car bodies in the 1920s was thus not only a reflection of the engineering ideal of mechanical silence, but also of a more widely embraced desire for convenient silence—the French phrase was *silence confortable*—that became increasingly important once long-distance travel became more common. Convenient silence, understood as encapsulating motorists by shielding them from exterior noise, was believed to ease the physiological effects of the noise that the car engine and chassis emitted despite efforts to make them more quiet.

ARISTOCRATIC SILENCE AS MARK OF DISTINCTION

"Speed is the aristocracy of movement, yet silence is the aristocracy of speed." This aphorism was formulated by Maurice Goudard, president of the French Society of Automotive Engineers in 1935 (Goudard cited by Kagan 1937: 14). It nicely captures the complexities embedded in the symbolism of the senses and the social hierarchies connected to this symbolism.

We have already explained that until about 1900 loud sound possessed a rather straightforward connotation of power, both within and beyond the engineering community. For twentieth-century mechanical and automotive engineers, however, the loud sounds generated by machines gradually came to be seen as the by-product of energy-absorbing friction and, as such, became noise, or unwanted sound. Culturally speaking, the symbolic links between loud sound and power were less easy to decouple. As many historians, soundscape scholars, and anthropologists have shown, Western societies have long been ordered in such a way that those in powerful and high-ranking positions possessed more rights to create loud sounds and make themselves heard than those in lower-ranked positions. To put it very simply: the king had his heralds, the general his gunfire, and the aristocrat his private orchestra, and nobody imposed a ban on these sounding activities (for an overview, see Bijsterveld 2008).

The elite, however, had more than the opportunity to make sounds that impressed others; they also had the financial wherewithal to build secluded spaces of silence such as castles and cloisters, and later luxury villas in bourgeois residential areas. Might, then, implied control over auditory space. The flip side of such social privilege was that those lower in the social hierarchy had to be quiet in the presence of the higher ranked, both literally and symbolically: monks had to be silent in the "presence" of God; servants and slaves had to be silent in the presence of their masters, women in the presence of men, and children in the presence of adults. And when the lower classes produced loud sounds, these were easily interpreted as signs of upheaval, social disturbance, or revolution (Burke 1993, Schafer 1994 [1977]; Smith 2004; Attali 1985; Thompson 2002; Picker 2003). As Peter Burke has shown, however, from the sixteenth century onward, the symbolism of sound became more complicated as the ability to keep silent—to hold one's tongue when necessary—came to be seen as a sign of self-control, civilization, and refinement (Burke 1993). Simply broadcasting oneself at each and every occasion gradually became less chic, and silence subtly grew into a mark of distinction: a bourgeois ideal with a touch of aristocracy.

Analyzing car manufacturers' interwar ad campaigns with this symbolism of noise and silence in mind clarifies that the automotive industry advertised "silence" in a way that not only did justice to mechanical and convenient silence, but also tapped into several varieties of old and new aristocratic silence. Maurice Goudard's aphorism from 1935 underlined that the speed of the elitist sportsman-motorist went very well with a silent car: mechanical silence and aristocratic speed were by no means mutually exclusive. That Goudard felt compelled to spell out this connection also suggests, however, how common the symbolic links between speed, societal dominance, and loud sound had been. Decoupling them required work. Alfa Romeo promised that its new six-cylinder landaulet would be "flexible, powerful, silent," explicitly associating power with silence (advertisement, Alfa Romeo 1922). Manufacturers did not rely on words alone to create convincing links between power and silence, however. They also used images. One of these was that of the panther. The panther is "courageous, supple, agile and strong," trumpeted a 1927 ad for the Lincoln (figure 2.6). "She is able to approach without noise, with an extreme slowness, and then suddenly, she jumps forward with lightning speed and always silent" (advertisement, Lincoln 1927; see also figure 2.7).

Figure 2.6
Lincoln advertisement (1927)
Source: Advertisement, Lincoln 1927.

Figure 2.7
Brampton advertisement, detail (1927)
Source: Advertisement, Brampton 1927.

Two other aristocratic symbols of silence used were that of a queen and an ancient goddess of silence. In an ad created for Panhard cars in 1920, a big hand is shown rising from a car, holding five playing cards, each signifying one of its featured qualities: elegance, speed, silence, solidity, and simplicity. Silence, designated by the queen of hearts, is shown in the middle (figure 2.8). A 1929 Delage ad used the image of a classic goddess of silence holding her index finger over closed lips, with that of a roebuck—signifying speed—and a panther, expressing suppleness and strength, below her (figure 2.9; see also advertisement, Delage 1928). The goddess refers either to Angerona, a Roman goddess with a variety of attributed meanings, or, less likely, Hesykhia, a Greek spirit personifying silence and quiet. In the Roman pantheon, Angerona was the deification of fear and anxiety, relieving men from such feelings. When shown with sealed lips, or with a finger near her closed lips, she signified keeping the secret of a mystery surrounding Rome's name, or secrecy generally. The Freemasons later adopted her image as an icon

Figure 2.8
Panhard et Levassor advertisement (1920)
Source: Advertisement, Panhard et Levassor 1920.

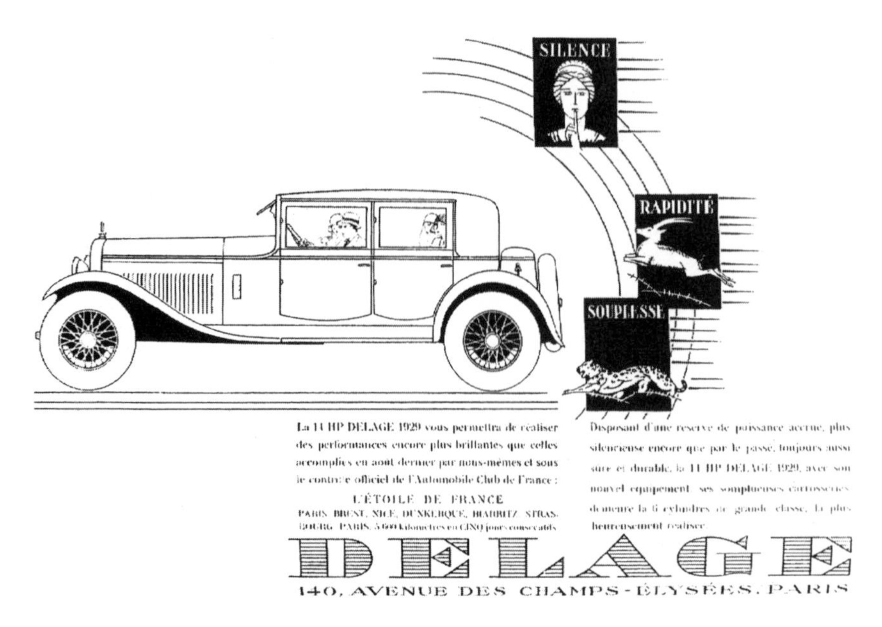

Figure 2.9
Delage advertisement (1929)
Source: Delage 1929.

of secrecy and silence. The Greek spirit Hesykhia referred to stillness, but much less is known about her iconography.[8]

Whether as queen, goddess, or spirit, silence is represented through monarchical or divine femininity here. It combines the symbolism of subordinated female silence with a high-ranked, aristocratic, distinguished silence, shown along with more "masculine" characteristics such as speed and strength. The use of such icons expressed the rise of aristocratic silence in the automotive industry, also nicely captured in a statement made by H.-G. Laignier, vice president of the French car agents association, in 1929. He recalled that some twenty years earlier "automobiles had to go as fast as possible, and they had to make a lot of noise. When you drove through a village you deliberately opened the exhaust. Today, preferences have changed: we want to hear a faint whisper, like the rustling of silk" (Laignier 1929: 604).

The phrase "rustling of silk" clearly refers to the abundant luxury of the rich. It is important to note that despite the investment necessary

8. See http://academialodge.org/article_silence.php (accessed August 17, 2012). We would like to thank classicist Marietje Kardaun and expert on antique and medieval philosophy Joke Spruyt for referring us to this source and providing additional explanation.

for delinking power and high volumes of sound, the relative silence of limousines and electric vehicles already possessed the connotation of *luxurious* aristocracy in early automobilism (Mom 2004: 25, 103, 247). As automotive historian David Gartman has explained, the "luxury classics, because of superior engineering and careful hand-fitting, were mechanically tighter and drove more smoothly. Their engines ran quietly, their transmissions shifted effortlessly and their brakes functioned at a touch, creating refined, relaxed driving experience befitting the ostentatious ease characteristic of the upper-class habitus" (Gartman 2004: 173). Sociologist Mimi Sheller has similarly flagged a historically developed alliance between a smooth and silent ride and ideas of luxury, privilege, and wealth (Sheller 2004). The Rolls-Royce was the archetype of such luxury cars, with one ad claiming that it was "as silent as its shadow" (figure 2.10). This ad expressed an old-style

Figure 2.10
Rolls Royce advertisement (1935)
Source: Roberts 1976: 140.
Courtesy: Rolls-Royce Motor Cars.

aristocracy and societal hierarchy: servants were the silent shadows of their masters par excellence. In 1923, Henri Petit praised the silence of the new 20 HP Rolls-Royce. As long as its speed did not exceed 70 kilometers per hour, it was as if "you are padded in cotton wool, . . . without the faintest noise, without the slightest vibration" (Petit 1923a: 168).

Alfa Romeo's sales department assumed that this would be the "dream of *any* motorist" (advertisement, Alfa Romeo 1922, our italics). And indeed, from the 1920s onward, aristocratic silence spread as an ideal across a wider array of car types. An example of the latter is Charles Weymann's way of marketing. He assured potential customers that "the silence of the Weymann body satisfies the most delicate ear," and presented this slogan with a drawing of an oversized ear behind a luxurious four-door sedan (figure 2.11). Delicate hearing was thus seen as a sign of civilized refinement. In 1926, the *Bulletin de l'Automobile Club de France*, the magazine of the French automobile club, announced the "Death of the Torpedo"

Figure 2.11
Weymann advertisement (1924)
Source: Advertisement, Weymann 1924.

(Charles-Faroux 1926: 25). The closed sedan now should be *the* choice of the stylish driver (Cartoon La Mode 1927).

When manufacturers such as Renault, Citroën, and Mathis started to produce sedans for the middle class, they labeled their sound as the "royal luxury of silence" (Charles-Faroux 1929b: 578; see also advertisement, Weymann 1929). By doing so, they stressed the new owners' refined taste. In addition, they visualized their novel standing by using aristocratic animal emblems such as the panther and the white swan, an icon of tranquility employed by Renault, Peugeot, and Citroën, the latter of which made the swan the symbol of its *moteur flottant* (figure 2.12; advertisement, Renault 1926b; Maillard 1932). In the United States, manufacturers used similarly elite symbols. Body by Fisher, for instance, used a Napoleonic coach as its visual icon (White 1991: 61). These labels promised middle-class motorists that the closed sedan would bring aristocratic silence within reach, as a sonic sign of social standing.

Figure 2.12
Peugeot advertisement (1928)
Source: Advertisement, Peugeot 1928b.

The success of the closed body coincided with another, related shift in American advertising: "Instead of the former emphasis on the reliability and performance of the car's mechanical elements, now it highlighted the pleasure and psychological benefits of driving it" (Bardou et al. 1982: 118). In France, manufacturers did the same. The French Society of Automotive Engineers, for example, invited H.-G. Laignier to acquaint automobile engineers with the latest marketing strategies. In a 1929 lecture, he declared the automobile to be "an object that rapidly becomes obsolete" (Laignier 1929: 605). Annual novelties in styling or accessories pushed new cars out of fashion within two or three years (Laignier 1929: 599–607). Marketers had to stimulate these fashion cycles and to arouse the customers' desires by exploiting all human emotions. The vanity of the motorist was the true engine of sales, as fashion nourished "the need of the elites to distinguish themselves from the masses, and the need of the masses to copy the elites" (Laignier 1929: 602). The distinguished silence of elites was one of the things to be copied. But how silent was automotive silence by that time?

AN UPLIFTING AND DISTANCING EXPERIENCE

As early as 1925, the year the decibel was introduced as unit for measuring "noise,"[9] specialists from Bell Laboratories explained the intricacies of noise production and measurement at the meetings of the US-based Society of Automotive Engineers (SAE) (Snook 1925). Nine years later, measurements of interior car sound levels forced American engineers to admit that Weymann's fabric body was quieter than the wood-steel and steel bodies produced by Fisher and Budd (Charles-Faroux 1934: 89–90). Soon afterward, the former Citroën engineer Charles Brull lamented that the mass-produced all-steel bodies still caused serious noise problems (1935: 261).

The quest for the silent body, Henri Petit and René Charles-Faroux stressed in tandem, thus had to go on (Charles-Faroux 1933; Petit 1934: 236–39). An "absolute silence" had been promised (advertisement, Manessius 1925). Yet car interiors were not silent, at least not in terms of contemporary noise measurements. An American study, published in 1937, claimed that the intensity of sound in the interior of a car going 60 miles per hour was 82 decibels. The publication did not disclose any information about

9. Today, acousticians define "noise" as "unwanted sound," distinguish noise from sound proper, and consider the decibel a unit for measuring "sound levels."

the make or type of car that had been measured, but given the American context and the year of publication, it was probably a wood-steel or steel body. Its sound was 2 decibels louder than the noise produced by a large symphony orchestra at that time (Critchfield 1937: 369). In 1938, Joseph Bethenod, vice president of the French Society of Automotive Engineers, published a sharp critique of the exaggerated claims of car manufacturers. "Each year, at the Salon," he inveighed, "we hear the ever-recurring chorus of engines running faster, with higher compression, more silent gears, etc., etc." Yet the cars coming off the assembly line did not come up to the mark (Bethenod 1938: 444).

Clearly, the interior sound of closed sedans did not live up to the expectations raised in the silence ads run during the interwar years. Given the complaints and measurements just presented, it is actually surprising that some automotive journalists testified to having had tranquil driving experiences in closed-body cars (see section 4), even in middle-class and cheap steel-body cars (Petit 1923b; Cazalis 1929). Placing these "earwitness" reports in a broader auditory context may be helpful in understanding the journalists' enthusiasm, however. A sound intensity of 82 decibels at 60 mph (nearly 97 kilometers per hour) in a car's *interior* is not likely to be perceived as "silent" by a test driver today, as the maximum *exterior* sound level a car is allowed to emit is 74 decibels.[10] Yet the 82 decibels of interior noise must have been *relatively* silent for those who had to live with cars at a time when an important British recommendation, published in 1937, for the maximum sound level of vehicles (cars and trucks) was 90 decibels for new vehicles and 100 decibels for used ones (Bijsterveld 2008: 121).[11] This relative silence becomes even more evident if one acknowledges that expressions in terms of decibels are based on a logarithmic scale. Thus it is comprehensible that in the 1920s and 1930s some test drivers were able to perceive "tranquility" in closed cars.

That automotive journalists still considered improvements in body noise control necessary can be understood by taking another sensorial context into account: the relationships between the auditory, kinesthetic, and visual experiences of driving a car. This requires a more lengthy explanation.

10. Council Directive 92/97/EEC of November 10, 1992, *Official Journal of the European Communities*, 19.12.92, No. L 371/1. This maximum, required for new cars, is measured at 50 kilometers per hour, under particular driving conditions.

11. At that time, the British committee expressed sound levels in "phons." A phon expressed the equivalent *loudness* level for a particular *sound intensity* level measured in decibels, with 100 phons (as defined by the International Committee on Acoustics in the same year) being the equivalent loudness of 100 decibels within a particular range of frequencies (Bijsterveld 2008: 106). Later, the decibel became the shorthand unit for expressing sound levels.

Let us return to the early days of the automobile for a moment. One of us, the historian of mobility Gijs Mom, has shown that a French-Belgian group of literati-motorists, active in the years prior to World War I, connected the perception of speed while driving to the sensation of experiencing the world from an uplifted, privileged, and distanced position, as in flight (Mom 2011; for the experience of driving as flight see also Desportes 2005: 247). One of the writers contributing to this group's style was Marcel Proust. In a story published in the Parisian newspaper *Le Figaro*, in 1907, Proust described the illusion a passenger experiences when approaching two adjacent towns—a description that became famous among literary scholars. The towns' church spires are moving toward the passenger, while he is remaining in place: the inversion effect (Nantet 1958; Danius 2001, 2002). Even though early railway passengers reported similar experiences (Schivelbusch 1977), the inversion effect became especially popular in writing about the car.

It is less widely known, however, that Proust also elaborated on the sensation of sound and vibration. For instance, he compares the chauffeur, clad in a long coat, with a "nun in the service of speed" as well as with an organ player changing registers, and has the "I" in the story hear music in the sounds of handling the gearbox (Proust 2009 [1907]: 245). Proust was not the only writer to do so. Rudyard Kipling had this to say about his chauffeur: "He improvised on the keys—the snapping levers and quivering accelerators—marvelous variations, so that our progress was sometimes a fugue and sometimes a barn-dance" (Kipling 1904: 137). In these cases, sound and vibration were similarly important to the experience of perceiving the world "from above," as was sight: an organ player both sonically and visually dominates the congregation.

The French-Belgian group used a less innocent vocabulary, however. Writers including Paul Adam, Guillaume Apollinaire, Tristan Bernard, Cyriel Buysse, Henri Kistemaeckers, Valerie Larbaud, Maurice Leblanc, Maurice Maeterlinck, and Octave Mirbeau had a filmic, Futurist-inspired style of writing, full of masculine aggression, which, in its disdain for common man, expressed the attitude of the sportsmen's automobile club rather than the petit bourgeois touring club. Octave Mirbeau's popular book *La 628-E8*—its title refers to the license plate of the author's CGV (*Charron, Girardot et Voigt*)—is a remarkable example. Critics have praised the novel for its deconstruction of linear narrative into "a succession of Impressionist tableaux," rendering an "atomisation of the senses" in "a new poetics, of speed, movement, a novel way of describing the world *from within*" (Ziegler 2007: 173; Roy-Reverzy 1997: 263, italics in original). In one of the trips Mirbeau reenacts, the plain appears "moving," another example of inversion, while the road behind the car is littered with "broken vehicles and dead beasts" (McCaffrey 1999 [1989]: 127). The car is depicted as a similar vehicle of aggression in *Le troupeau de Clarisse* (1904), written by Paul Adam, and in *A.O. Barnabooth* (1913), by Valerie Larbaud. Larbaud captures an outing by a few superrich men who, just for the fun of it, decide to drive into a populated area of Nice to terrorize its residents. As Mom emphasizes, these and other novels from this French-Belgian group of writers display the contempt of some wealthy, higher-class people for those of lower social rank at precisely the moment that the petit bourgeois started to consider the car an appropriate means of transport for themselves (Mom 2011).

While these writers articulated processes of social distinction through the experience of speed, inversion, sonic, and visual dominance in *open* cars, driving a *closed* sedan gave the sensation of disassociating oneself from one's environment a new twist. It was the physical encapsulation in steel and wood and the relief from that enclosure provided by glass windows and side mirrors that created a novel privileged position for the expanding category of middle-class drivers and passengers. It enabled them to engender a more elite way of looking at the world that Pierre Bourdieu calls the "pure gaze," "a quasi-creative power which sets the aesthete apart from the common herd" (Bourdieu 1984: 31). Two quotations may illustrate this connection between social distinction and sensory distancing through the closed sedan. The first one is from 1922. It is a fragment from an anonymous travelogue published in the popular American literary periodical *Scribner's Magazine*:

> Two of us were lately guests in a great town, and had a limousine at our command. We actually ended our stay without once rubbing elbows humanly with any of the people in the streets, shut away from our fellows in a glass box, lifted out of the very life we had come to live, as though we had been looking on at a movie. (Quoted in White 1991: 63)

The second quote comes from an article entitled "Automobile Philosophy," which appeared in the 1929 *Allgemeine Automobil-Zeitung*:

> Separated only by thin glass, you keep your distance from the outside world. With the closing of the car door a new stadium is reached, a different individual starts to live.... The automobile world presents itself through the windows of a sedan.... A new, exciting, and desirable world, full of variety, and more entertaining—a world that approaches from the outside, less harsh and imperious, but rather charming and humble. (König 1929: 17)

Both authors describe the experience of enclosed driving as keeping aloof from the world outside, physically and emotionally. They express relief at being freed from physical contact with other people—the people in the street. Their pure gaze is also a *cinematic* gaze that facilitates the aesthetization of the world beyond the car, rendering it more desirable and entertaining than the real world (for cinematic driving experiences see also Desportes 2005: 269). From the lifted position of the driver and passengers, life outside the car appears less dangerous, in lower position, and visually controlled.

As literary scholar Andrea Wetterauer has pointed out, such descriptions are reminiscent of the famous ethnographic observations published by the German sociologist Georg Simmel in 1903. Simmel showed that around 1900 city dwellers developed a more blasé attitude toward their fellow citizens in response to the acceleration and intensification of modern life. Handling the excessive number of encounters in public required a new, more distanced mode of "seeing" others—visually a form of neglect. Wetterauer claims that the car added a novel dimension to this desire to flee from the impositions of modernity and to create a distance between oneself and the rest (2007: 74, 85–92).

In some of the stories about the car, driving was imagined as flight itself, a fantasy that accompanied the driving experience as early as the turn of the twentieth century (Mom 2004: 38). In Ehrenburg's *The Life of the Automobile,* one of the protagonists experiences a drive in the car as a "maddening flight" and refers to having read that cars can "catch up" with "any swallow" (Ehrenburg 1931 [1929]: 239). And in 1932, the American landscape architect Wilbur Simonson used metaphors of flight to describe a US highway he designed: "[T]his broad paved highway will simulate in its flowing lines, the spiral curves, the horizontal and vertical transitions, and the banked turns of a fast transport aircraft in flight." (Simonson quoted in Davis 2008: 52; for the publication's date, see Davis 2008: 44.)

SAVING THE CINEMATIC DRIVE

Enabling a smooth flight through *car* design rather than road design was not something that happened overnight. First of all, the closed car did not put an end to all the problems created by wind and weather. Keeping the windows closed helped to shut out the vagaries of weather, yet forced drivers and passengers to breathe bad, vitiated air. Opening the windows, on the other hand, created drafts (Loewe 1930). In 1933, Fisher developed its famous no-draft ventilation system, allowing the inflow and circulation of fresh air while deflecting drafts. A year later, the German automotive trade journal *Kraftfahrzeug-Handwerk* published an article on how to seal the pedal openings with brushes (Anonymous 1934). It was as if the closed body triggered a new sensorial awareness among car owners, foregrounding small irritations they had not noticed in an open car. The insulation from the world beyond the windshield became a major task for automotive engineers.

A close examination of the Society of Automotive Engineers' journal clarifies how much effort had to be invested in constructing a smooth driving experience, and how all the driver's senses were taken into account (Mom 2008b). In 1922, the US National Automotive Chamber of Commerce (NACC) sent a questionnaire to 20,000 automobile owners about the characteristics they valued most. Ten percent of the recipients replied, putting "endurance" and "economy of operation" on top of their list, followed by "comfort," "price" and "appearance," while "speed" finished remarkably low in the ranking (Shidle 1922: 351). The outcome was remarkably similar to the results of the 1922 French questionnaire (discussed in section 4). Nonetheless, manufacturers and engineers did not follow owners' preferences slavishly. One of their reasons was that quick design changes raised production costs too much (Wolf 1933: 16). Comfort was seen as important, however, and, after some initial doubts about the comfort of American cars in comparison with European ones (Beecroft 1919), comfort was hailed as a special quality of the automobiles produced in the United States. An American engineer even stereotyped the average European motorist as a "speed-hound" who made his "noisy" engine roar at 4,000 rpm, whereas America was "the paradise of the lazy driver" (Johnson 1922: 306).

The US engineers acknowledged that the "demand for quietness" (Franzen 1928: 82) was especially intense in the luxury segment—noise was said to be "inversely proportional to price class" (Taub 1930: 723). Yet by the mid-1930s, the prime time of noise abatement campaigns, consumers had become so "noise conscious" (Anonymous 1934a: 37) that lower-priced cars were fitted with features that brought them closer to the "noise characteristics of the heavier, expensive car." One way of getting there—in addition to the interventions discussed in section 3—was adding absorption material that filtered out high-pitched noises, giving light cars a "heavier feel" (Prudden 1934: 267). Noise reduction, the engineers concluded at a symposium, "tends to reduce the feeling of vibration, even though the vibration itself has not been reduced" (Anonymous 1934b: 62).

Vibration, then, was seen as a crucial aspect of comfort as well. In the early 1920s, tire manufacturers had noticed that some customers deliberately underinflated tires in order to create a more comfortable ride at higher speeds, even if this increased tire wear (Hale 1923a; Hale 1923b; Lemon 1925). Tire manufacturers like Goodyear developed "balloon tires" with a stronger carcass, making them better able to withstand lower air pressure and enabling the tire to take up more of the energy of the total suspension. Engineers from the car manufacturing industry fiercely opposed this innovation, because it enhanced the risk of "shimmying," a heavy vibration of the wheel that compromised the stability of steering and driving, a "pernicious malady of mystery which is defying the cunning of near all car engineers." What's more, shimmying amplified noise emission (Burkhardt 1925: article title; Anonymous 1924c: 482). Yet within a few years almost all cars had balloon tires, and not long after, even superballoons (Lemon 1932).

This forced the automotive industry to begin research into the very elusive phenomenon of riding comfort. Some senior engineers distanced themselves from this line of research, claiming that customers did, in fact, not want the "very monotonous" and "extreme boredom" of "an absolutely smooth road," even if these customers were *said* to consider this situation as "ideal." In contrast, they averred, consumers actually loved the "heroic" vibration of the engine that "gave the impression of high speed." But most automotive engineers contributed to taming the automotive "adventure" in terms of its acoustics and vibration (Anonymous 1925c: 392; see also Horning 1925: 190; Hess 1924a: 82, 1924b: 543; Anonymous 1930a: 99, 101; Anonymous 1930b: 137).

Acknowledging that the American automotive industry had "no satisfactory yardstick to measure riding-comfort," but that "vibrations . . . are by far the greatest annoyance in a car," the Society of Automotive Engineers approached psychology professor Fred A. Moss to chair a Riding-Comfort Research Subcommittee in 1925. This committee placed individuals on a "vibrating chair," transforming their bodies into seismographs, and defined "fatigue" as an important sign of low levels of comfort. It also commissioned studies in which test subjects were asked to assess an impressive range of automotive qualities in terms of the sensation of motion, sound, sight, smell, and aesthetics—testifying to a genuinely multisensorial approach to the driving experience. The research showed that women considered "skidding" to be more objectionable than men, while both sexes preferred the closed car over an open one and felt that noise was disagreeable. It also found that unknown sources of sound caused anxiety (Moss 1930a, 1930b; Anonymous 1930c: 716; Brandenburg and Swope 1930; Moss 1932). Comfort, one engineer said, was a "state of mind," clearly affected by "mysterious rattles, squeaks and grunts" (England 1930: 70–71). It remained hard to define a "a good ride," however (Kindl 1933: 176), and researchers kept trying to tackle vibration by introducing engineering concepts such as "roadability" and "harshness," and having a specialized "ride engineer" join the ranks of the SAE (Muller 1930: 762; Hicks and Parker 1939: article title; Paton 1938: 313; see also Hess 1933; Tea 1934; Brown and Dickinson 1935; Jacklin 1936; Lay and Fisher 1940).

An even more pressing design problem, however, was the decrease in visibility that came along with the closed car body. The introduction of the windscreen had already transformed the visibility of the surrounding environment from a 360-degree panorama into a film screen on which only the *forward* part of the landscape seemed to be projected (for the rise of the "orientation frontale" see Desportes 2005: 312). Moreover, the driver's glance in the rearview mirror reduced and intensified the driver's side- and rearward gazes. A person's sight from the car was limited even further in cars with closed bodies, not only because of the surrounding metal but also because unsatisfactory ventilation caused the windows to mist up. Initially, the closed body was widely criticized for this loss of vision. Even when ventilation improved in the 1920s, drivers had to develop the special skill of focusing on what they saw in front of them, at a distance, while at the same time keeping an eye on the rest of the landscape, including the foreground.

Manufacturers attempted to convince future customers that their cars "permitted a view as perfect as that of a torpedo" or "with the greatest visibility" (for these quotations, see figures 2.13 and 2.14). Yet while single-unit construction enabled engineers to lower the body and streamline its shape with rounded corners and slanted windscreens, the increasing wind resistance and tire circulation resulting from higher speeds introduced new sources of noise. By the end of the 1930s, leading officers of SAE noticed that "appearance" was such a hot item that car owners "will accept a certain degree of discomfort" (Crane 1939: 141). At the same time, however, engineers openly started to fantasize about an "ideal car [that] should rapidly traverse an ordinary road in such a manner that the only indication of motion would be the sight of the passing landscape" (Hicks and Parker 1939: 1). This reveals why engineers had invested so much in sound and vibration: noise and harshness had to be reduced to keep both the endangered cinematic gaze and the ideal of flight alive. As a 1929 Peugeot ad suggested by displaying a movie reel over a fast-moving car, a filmic experience of the landscape was within reach in a closed sedan "sans bruits, sans heurts," or "without noise and shaking" (figure 2.15). With every bump engineered away, with acceleration ironed into a smooth and relatively silent experience, the car could increasingly be driven around as a rolling advertisement for the cinematic drive and the dream of flight.

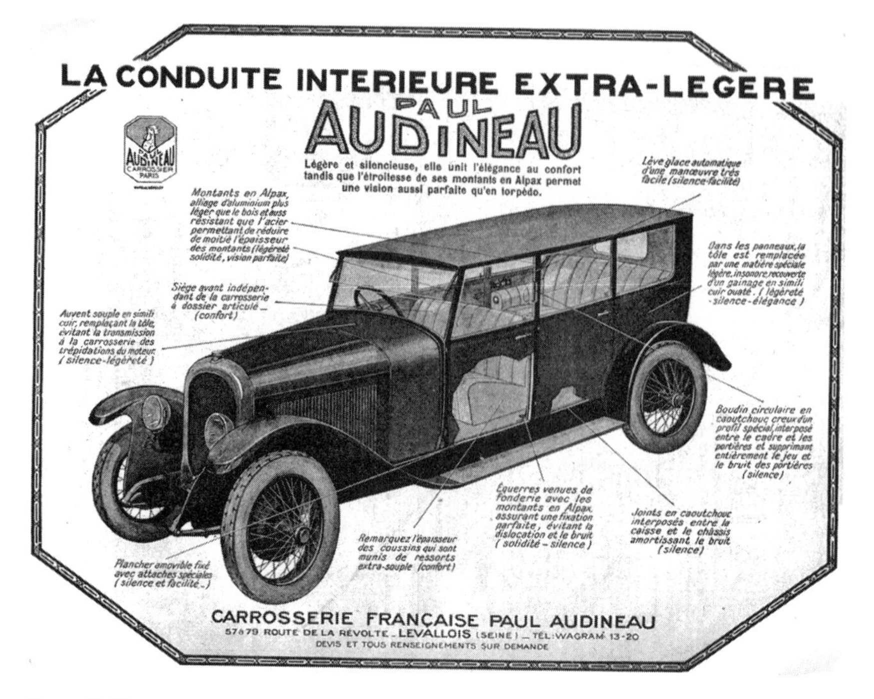

Figure 2.13
Audineau advertisement (1925)
Source: Advertisement, Audineau 1925b.

ORCHESTRATING THE CAR

In the process of closing its body in the 1920s and 1930s, the car developed into a room on wheels that enabled a celebration of the cinematic drive through a carefully designed *orchestration* of sound and vibration. The engineers who worked on mechanical silence and the manufacturers who advertised convenient and aristocratic tranquility selectively responded to test drivers' and consumers' need for long-distance travel without noise-induced fatigue. Yet all these groups together contributed to the paradox of the closed body car: it prolonged the driver's experience of visually dominating the road and its environment, despite the fact that it limited the driver's view, *because* early sound design dampened vibration and noise, thus contributing to the experience of driving akin to flight.

It is worth noting that, in fact, the chic silence increasingly desired by middle-class motorists remained largely a privilege of the wealthy. Measured in terms of decibel, the interior of the mass-produced closed sedans was not "quiet" at all. Yet the manufacturers of low- and medium-priced brands kept advertising their make's putative silence by

Figure 2.14
Manessius advertisement (1926)
Source: Advertisement, Manessius 1926.

applauding its exclusiveness. As late as 1935, Pontiac advertised the Silver Streak in this manner: "If your ideas of luxury run to fine fabrics exquisitely tailored, sparkling chromium fittings, and an adroit use of color, the jewel-box perfection of Pontiac interiors is sure to win your heart. And if you measure luxury by *comfort* Pontiac scores again! Deep cushions receive you... ample weight and special springs make your silent progress smooth and steady" (advertisement, Pontiac 1935). And Plymouth claimed about its cars that they were "scientifically insulated against engine vibrations and road noises." Only in a Plymouth, said another ad, do "you *really* relax" (advertisement, Plymouth 1937).

Clearly, the idea of a closed car that could be sonically designed in such a way that it would be a relaxing space for its driver and passengers had crystallized; and it was here to stay. The essay in illustrations in figures 2.16 through 2.29 in the appendix to this chapter testifies to the rise of the ideals of mechanical, convenient, and aristocratic silence in the 1920s and

Figure 2.15
Peugeot advertisement (1929)
Source: Advertisement, Peugeot 1929.

1930s. In addition it shows how these ideals kept popping up in marketing campaigns for cars after World War II, adding new meanings to them. In 1953, for instance, Fisher used a wonderful illustration of a forest with two Bambi-like deer, one of which dares to approach—with pricked-up ears—the interior of a car with a sound-insulating body made by Fisher. Fisher produced these bodies for the General Motors brands Chevrolet, Pontiac, Oldsmobile, Buick, and Cadillac.

One image Fisher employed in 1967 transforms the car's stillness into a "Quiet zone" for a mother and baby in the backseat. Here the convenient tranquility in the car acquires an additional layer of meaning: the car's silence brings not only relaxation and relief from sources of fatigue, but a quiet zone, the best possible protection from a noisy outside world—a world that increasingly defined noise as an environmental problem. In the quiet zone of the car even a child can find "sweet dreams." Also highly intriguing is a series of late 1950s Fisher ads that visualize the ideal of the smooth drive by presenting cars as flying objects. And finally, we would like to point out two pictures constructed for a Cadillac SRX ad campaign in 2011. This campaign promotes a "Library-Quiet interior," thus taking up the theme of elite silence again, even though it is in a bourgeois rather than an aristocratic version (for the sources, see the figures).

In the meantime, when drivers were touring and enjoying the landscape, they had started singing. Now that the car, as the German publicist Christian Bock explained in 1937, "acoustically isolated" the motorist "from the environment, whose customs and traditions forbid him to sing," the driver felt free to raise his musical voice. "Who ever saw a pedestrian sing on the streets of the city?" (Bock 1937: n.p.). "No one" is the answer the reader is supposed to give to his rhetorical question. But in his insulated space, the driver dared to sing. Soon the human voice would find a companion in the car radio.

Illustration Essay

Figure 2.16
Zahnradfabrik Friedrichshafen (ZF) advertisement (1934). This advertisement by the German gear-wheel manufacturer ZF promotes the "Höchste Laufruhe," or the highest running tranquility for its gear wheels, and nicely illustrates the engineering ideal of *mechanical silence* that acquired significance in the automotive industry in the 1920s and has remained important ever since.
Source: Advertisement, ZF 1934.
Courtesy: ZF Friedrichshafen AG.

Figure 2.17
Timken Axles advertisement (1932). In this ad, we see a person, dressed like a doctor in a white coat and wearing spectacles, listen to a car's transmission with a stethoscope. Timken Axles are so silent that "[n]ot even a stethoscope could catch the pulse of this artery of power transmission." Just like the previous figure, this ad expresses the pride of *mechanical silence* in the automotive industry. At the very same time, it flags the importance of the practice of listening to machines in order to monitor their functioning and diagnose potential problems, a theme we will discuss in the next chapter.
Source: Advertisement, Timken Axles 1932b.

Figure 2.18
Timken Axles advertisement (1932). This ad promotes the *mechanical silence* of the Timken Axles "Worm Drive" once again, and shows the cultural specificity of the ways in which the car-manufacturing industry and marketers tried to evoke silence on paper. Here, it is a world covered with snow and the act of gliding that creates the association with silence. It is not a boring or powerless silence, however, but an exciting, speedy, joyful, and youthful one. Pointing to the silence of one's product might create the worry that it had no power, and silence might even be associated with death. Yet this picture expresses both liveliness and an enduring silence: "It coasts, and coasts . . . and coasts silently."
Source: Advertisement, Timken 1932a.

Figure 2.19
Renault advertisement (1928). The comfort of "silence" in an automobile became something to search for when drivers increasingly began to use the car, and notably the closed car, for family outings and touring across the country instead of showing off their adventurous and sporty character. The noise-induced fatigue people struggled with when driving long stretches became an issue in the automotive trade literature, and the *convenient silence* of a well-manufactured car body was the answer. This ad nicely captures this modern-day "grand tourism" by allowing the spectator a vista on both the car and a lovely countryside, and by underlining Renault's "suppleness" as well as its "silence" among its many "stylish" comforts.
Source: Advertisement, Renault 1928.

Figure 2.20
Body by Fisher advertisement (1935). We see a young girl, about four or five years old, sitting at the doorstep of a house in snowy weather: "Safe and sound, this personable young lady has been delivered by motor to her doorstep, in a comfort and safety which to her are a matter of course." That the car is not "drafty" (important in such cold weather) is something she doesn't give much thought, nor does she "pay much attention to the good solid thud of a door swinging shut, or consider the superb Fisher craftsmanship which accounts for that safety and ruggedness," even though she "does like to snug down on the cushions." But her "elders" do notice, and know they better buy a car with Body by Fisher. Clearly, the solid thud of the door signifies the sound of safety. It expresses an important shift in advertising car sound design, in which the body's *sound*—combined with other qualities affording cocooning, such as cushions—*not only stood for comfort and convenience, but also for a robust safety*. The girl has been brought home "safe and sound," and we see how "sound and safe" have been brought together.
Source: Advertisement, Body by Fisher 1935.
Courtesy: General Motors LLC.

Figure 2.21
Lincoln advertisement (1926). "Harmony" is the title of this ad and the name of this particular Lincoln make. It immediately rings the bell of sound. The ad highlights that everything in the Lincoln runs smoothly, that it is a "genuinely rolling home," "silent and supple" and that the car radiates "power and safety," another take on what we have coined the ideal of *convenient silence*.
Source: Advertisement, Lincoln 1926.

Figure 2.22
Panhard and Levassor advertisement (1931). In the interwar years, car advertising promoted not only mechanical and convenient silence, but also *artistocratic silence*. Being able to keep quiet had been an age-old sign of distinction. "Distinction" is even in the title of this ad. The Panhard on display combines "mechanical refinement" with four "silent speeds" (or gears), which results in a "harmonic ensemble of rare distinction."
Source: Advertisement, Panhard et Levassor 1931.

Figure 2.23
Hotchkiss advertisement (1930). This ad elicits both the silence of gliding, a trope we have already come across in the ad with the children gliding through snow, and the elite and elegant sport of sailing. It is therefore an example of expressing *aristocratic* silence. Like a sailing yacht, a Hotchkiss is able to accelerate "without noise."
Source: Advertisement, Hotchkiss 1930.

Figure 2.24

Body by Fisher advertisement (1953). A car—Chevrolet, Pontiac, Oldsmobile, Buick,or Cadillac—with a body made by Fisher is so "quiet," this ad seems to say, that even an animal as shy as a roe might approach it with curiosity. The ad illustrates that creating a quiet car interior remained an important goal in the automotive industry after World War II. In addition, the text of the ad introduces two new themes. First, it mentions the presence of other "traffic" and what that may mean to the sonic environment of drivers and passengers, a topic we will elaborate on in the next chapter. Second, it explains the technology used by engineers to develop automotive tranquility. It also returns to the familiar ground of aristocratic silence though, as Body by Fisher still used the visual icon of the Napoleonic coach (see the box displayed in the left corner at the bottom of the ad). Here are some quotations from the ad that show the old and new themes: "There are times when even the slightest noise can spoil the picture—such a picture, for instance—as your motoring comfort. That's why Fisher body sound engineers go to such lengths to protect you against the irritation of the many little noises which are bound to result from a sizable object like a motorcar, moving over rough roads and through traffic." The sound engineers' "search for silence continues—at the special Fisher Body sound laboratory. Right now, for instance, they are even making use of the latest binaural sound equipment in their work. This consists of double microphones, attached to a tape recorder on the rear seat of a test car, which register sound exactly the way a person with normal hearing catches it."

Source: Advertisement, Body by Fisher 1953.

Courtesy: General Motors LLC.

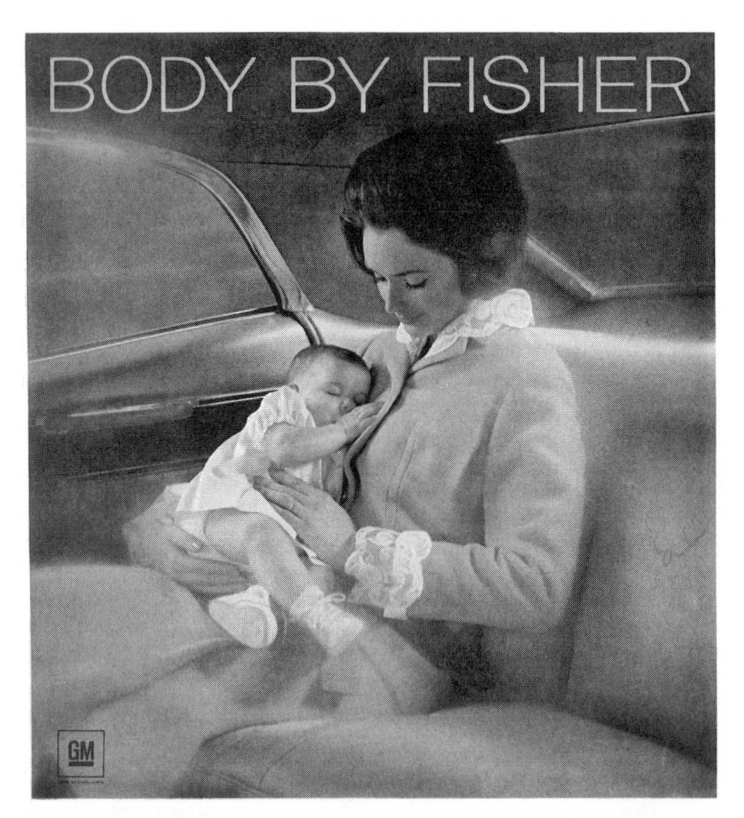

Figure 2.25
Body by Fisher advertisement (1967). This 1967 ad is like a remake of the ad from 1935, the one with the girl delivered "safe and sound" by motor. Here we see a mother and baby *inside* the car instead of a child *outside* the car. Yet again the theme is the safety expressed by the sound of a car, even though it is silence more than the solidity of sound that is advertised in this case: "Sweet dreams, little one. Cradled ever so gently in Fisher Body elegance. With soft layers of insulation to soak up sound. Thick cushions of rubber to shush vibration. Plush fabrics and carpets to add a final quiet note. Rest easy."
Source: Advertisement, Body by Fisher 1967.
Courtesy: General Motors LLC.

Figure 2.26
Body by Fisher advertisement (1958). "You're snug as a bug in a rug in the new 'sound barrier' Body by Fisher," this ad says. Its title recalls the trope of acoustic cocooning, and its picture shows a smoothly flying car amid heavy city traffic. As we have explained in chapter 2, the automotive industry conceptualized the reduction of noise and vibration as contributing to a smooth, cinematic, and "flying" drive. Similarly, as we will show in chapters 3 and 4, car radio was thought to give "wings" to the car (see figures 4.9 and 4.10). By using the notion of "sound barrier," the ad also evokes an association with the Concorde airplane, the supersonic passenger jet airplane that was under construction from 1955 onward, and was to be able to break through the sound barrier. It was an icon of modernity at the time, until it actually started flying in 1969. Eventually, the noise it created, the accidents it caused, and the expenses it required made the general public turn against the airplane itself. The association between Body by Fisher and the flying Concorde is speculation, though. This is what the ad actually claims while showing a flying car: "Every Fisher Body has snug-fitting joints, taut seams, 23 different kinds of sealing and insulation. Result: moisture and dirt are sealed *out*, silence is sealed *in*, rattles are shrugged *off*."
Source: Advertisement, Body by Fisher 1958b.
Courtesy: General Motors LLC.

Figure 2.27
Body by Fisher ad (1958). This ad is another version of the "flying" car, now underlining—while picturing age-old pyramids—that Fisher bodies are "built to stay silent for years." Moreover, the ad explains the differences in measurement of interior car sound with an "electronic ear" before and after the introduction of the twenty-three ways of sealing and insulation also mentioned in figure 2.26. Note that the "after" picture displays a more smooth visual representation of sound than the rough lines of the "before" picture.
Source: Advertisement, Body by Fisher 1958a.
Courtesy: General Motors LLC.

Figure 2.28
Cadillac SRX advertisement (2011). This ad illustrates the continuing efforts in car marketing to sell interior car sound as an important automotive quality. It also shows that Cadillac plays the card of a "Library-Quiet Interior." In this case, we see a classic library with traffic lights at its center by way of stressing the tranquility of the car amid a crowded road. Eliciting silence by picturing a library is historically interesting, as libraries are only silent spaces since the introduction of silent reading in the tenth century (Manguel 1999 [1996]: 57–70). The symbolic history of silence also shows a long tradition of associating scholarship with silence. Early modern universities, for instance, secured silent working environments for their professors in several ways (Otterspeer 2000: 297; Wiethaup 1966: 122). Today's libraries, however, are not as silent as their nineteenth- and twentieth-century predecessors (Mattern 2007). It is no coincidence, then, that Toyota Avensis has claimed that its interior is *more* quiet than a contemporary library. See our YouTube essay in chapter 5, section 2.
Source: http://cargocollective.com/raffa/CADILLAC, retrieved September 5, 2012.
Courtesy: General Motors LLC.

Figure 2.29
Cadillac SRX advertisement material (2011). This illustration explains an advertisement strategy by Cadillac and plays with the same theme as figure 2.28: the library-quiet interior. It is about providing drivers with a sunshade for their cars' windscreen on which a photo of a library has been projected. As the material explains, the sunshade is meant to "[r]aise awareness about the 2011 Cadillac SRX silent interior, called Library Quiet."
Source: http://cargocollective.com/raffa/CADILLAC, retrieved September 5, 2012.
Courtesy: General Motors LLC.

CHAPTER 3

Humming Engines and Moving Radio

Encapsulating the Listening Driver

LISTENING TO OLD IRON

In the mid-1990s, the Ghanaian taxi driver Kwaku struggled with his imported secondhand Peugeot 504. As the anthropologists Jojada Verrips and Birgit Meyer observed, Kwaku's car "looked like a heap of scrap-iron beyond repair." Yet repairing it was exactly what Kwaku was up to since he needed the car for making his living as long-distance taxi driver. Like many other taxis in Ghana, the Peugeot had the words "God never fails" written on its body. The one never failing, however, was Kwaku himself: he spent all his time and money in keeping his car rolling on Ghana's many "defective roads." This fueled the anthropologists' interest in Ghanaian car culture, in which men like Kwaku "adjusted" and "tropicalized" cars from the West, repairing them "with a minimum of tools and an often discouraging lack of spare parts" (Verrips and Meyer 2001: 155–56). To their surprise, Kwaku "constantly referred to how he used his senses (especially his ears and nose) to find out how his car was functioning." Similarly striking was the way he talked about his sensory skills, saying things like "I felt a smell" and "I heard a scent" (Verrips and Meyer 2001: 183).

If Kwaku's idiosyncratic repair practices and those in today's tool-stuffed garages of the West seem miles apart, such practices were still quite common in Western countries in the first half of the twentieth century. This chapter shows that diagnostic listening to car engines was considered an everyday duty of drivers in the first decades of the twentieth century, but gradually came to be seen as something that should be the exclusive expertise of mechanics. This was especially true in Germany, where car repair evolved into a protected, guild-like profession in the mid-1930s, but an analysis of European manuals for motorists published between the 1930s and the 1970s provides evidence of such development beyond Germany as well, even though the backgrounds behind the shift may have been different.

Intriguingly, this process of delistening to the engine—as pertaining to drivers—was paralleled by an increased listening to the car radio, a process that we will follow from the 1920s to the 1970s. As we will show, the meaning of car radio shifted over time and in telling ways. Whereas early car radio was sold as a companion to the lonely driver, in the 1960s it came to be defined as a sonic assistant helping drivers to cope with their *lack* of solitude on the road: to musically control their temper in situations of traffic jams and badly behaving fellow drivers. The listening driver was thus encapsulated in new ways: not only mentally detached from the sound of the engine, but also guided by the mood-improving sound of his car audio equipment. So while the previous chapter discussed the early initiatives for shielding drivers and passengers *acoustically* from the sounds of their cars, and for selling them the new "silence" with a rich set of connotations, this chapter is about how drivers ought to listen to their cars. It tells how changing professional practices among mechanics and advertisers' strategies to have the car radio accepted, fostered a tuning of drivers' ears to radio sounds that had to make them feel safe and well tempered—sounds that filled their cars and *metaphorically* shielded them from the dangers of the road.

As in chapter 2, trade journals for automotive engineers and special interest journals for automobilists have informed our claims significantly. Particularly the periodicals of automobile clubs are rich historical sources, as they contain hundreds of letters from motorists in which they describe technical problems—often mentioning their listening practices. We have also examined journals for auto mechanics and garage owners and considered handbooks for mechanics, chauffeurs, and car owners. We started with a focus on Germany, where car mechanics secured a particular jurisdiction and position for their craft, while the United States, where this did not happen, proved to be an interesting counterexample. Driver's manuals published in European countries other than Germany helped us to widen up our coverage of automotive listening practices. As we have already indicated in chapter 1, Philips was a dominant player in car radio manufacturing, and its archives clarify how the Dutch electronics company learned from the introduction of car radio in the United States to smoothen the road for radio in Europe. The archives of Philips's German competitor

Blaupunkt enabled us to put the Philips design and advertising strategies in a wider perspective.[1] A study of radio company archives and of advertisements and articles in radio journals, in addition to the automotive sources we have already mentioned, makes clear how Philips and other manufacturers embedded their radios in European car manufacturing and driving practices, and how motorists learned to listen to these audio technologies.

As we discussed in the previous chapter, early American studies on driving experiences showed that unknown sources of noise often worried the driver. A heightened level of attention when being confronted with unidentifiable sources of sound may be a human condition, as unknown sources are potentially dangerous. The Chicago psychologist Richard R. Fay has argued "that the primitive function of the auditory system is not for vocal communication, but 'to obtain information on the identities and locations of the objects in the immediate world that produce or scatter sound'" (quoted in Plack 2010: 164). Other experts in auditory perception have claimed that test subjects' attention is enhanced when they are presented with a "spatially non-predictive auditory cue" shortly before a task in auditory location (Plack 2010: 260). What is "non-predictive" or "unidentifiable" to whom, however, happens to be historically variable in intriguing ways—at least in the automotive world, as the next sections will show.

MYSTERIOUS NOISES: THE DRIVER'S EXPERIENCES

It was a hot summer day in 1950 when a Bavarian driver decided to bring his car to a nearby repair shop. He complained about hearing an inexplicable slapping noise when driving, and worried about the technical problem this might indicate. The mechanics at the garage started to search for the source of the noise. They checked the carburetor and the ignition in particular—these often caused trouble. They could not find the slightest malfunction, however. That same afternoon, the motorist returned to pick up his car. The mechanics asked him to describe the slapping sound in more detail: when and where did he hear it exactly? "Each day I drive down a long, straight street lined with poplars," he answered, "and it is there that I always hear the slapping noise" (Anonymous 1950: 353). The mechanics

1. In the mid-1950s, Philips became a major player in the European car radio market (Weber 2008: 143). Twenty years later, Philips dominated this market, together with Blaupunkt (Fesneau 2009: 367). The Philips archives offer more information than the Blaupunkt archives about the deliberations *behind* the companies' marketing and advertising strategies.

doubled up with laughter. They instantly understood what was happening: the driver had been hearing the echo of the car itself reflected back by the trees and their foliage.

In this case, the mechanics took the driver's ability to understand his car's sounds seriously for only half a day. At the end of World War I, however, the situation was quite different. German automotive experts expected a spur in the postwar demand for small, light, economical, and easy-to-drive cars now that the war had proven the car's usefulness and reliability (König 1919; Ledertheil 1919). Such cars, they presumed, would be bought by "self-driving" tradesmen and other middle-class people with a reasonable level of automotive knowledge (Hessler 1926: 7). The automotive literature of the day reflected these assumptions. Many an instruction book addressed both professional chauffeurs and self-driving automobilists, and periodicals for mechanical engineers and ordinary motorists might still publish the same articles (Ostwald 1921; Parzer-Mühlbacher 1926; Hacker 1932; Martini 1938).

What, then, did the self-driver need to know and to do what the chauffeur had previously known and done? Apart from driving, the chauffeurs' main duties had been maintenance and repair work. Only major engine jobs requiring specialist equipment were to be left to repair shops (Martini 1922: 13; Möser 2009). Chauffeurs therefore needed a high level of technical knowledge and, remarkably, a good sense of hearing. "The ear should rigorously register the finest deviance of the engine sound," an automotive expert claimed (König 1919: 12). While driving, the chauffeur had to listen carefully, and in case he detected even the slightest dissonance, he had to look for its source and decide whether it required immediate repair (Küster 1907: 10; Martini 1922: 207). It was only through experience that the chauffeur could develop such a fine sense of the engine's rhythm and pitch. For a proper diagnosis of engine problems, however, he needed a good theoretical understanding of automotive technology (Küster 1907: 10).

The self-driving car owner would find similar advice. In one of the earliest handbooks for drivers without chauffeur, a handbook appearing in as many as thirteen editions between 1904 and 1930, the motorist was urged to "listen to the desires of his engine" (Schmal 1912: 10). Other manuals stressed the importance of recurrent listening to the sound of the machine (Hacker 1932: 62). Once the motorist noticed discord, he was to drive carefully and "open up his ears" (Hacker 1932: 83). Drivers were instructed to avoid unnecessary noise because they would only notice malfunctions "plainly and early" when the engine ran "as quietly as possible." Too much noise might also harm drivers' sense of hearing, and negatively affect their sensitivity to what the machine needed (Hessler 1926: 217, 203). In general, all motorists could train their ears:

> With growing experience and habit even the beginner learns to focus his attention on... his own car, without being distracted from the road.... He will soon notice that every engine and every car has its own pace and that even the slightest technical problem alters this lovely rhythm.... A knock or rattle of the engine, a crunch of the chain, a rattle of a bolt will indicate the spot where the car needs maintenance, and he will do well to follow the slightest hint to repair malfunctions in time before they grow worse. (Küster 1919: 304)

As this handbook for automobile maintenance suggests, motorists needed time to get to know their car and learn to understand what the engine said to them. They had to acquire listening skills that enabled them to distinguish between the sounds of a properly running vehicle and those indicating problems—a mode of listening that might be called "monitory listening." Monitory listening is "the kind of auditory surveillance that scientists, engineers, and physicians employ in order to check the proper functioning of instruments, machines and patients' bodies." (Pinch and Bijsterveld 2012: 14; see also Mody 2005; Bijsterveld 2006, 2009). It is different from "diagnostic listening," a mode applied in medicine and engineering to identify the pathologies and problems behind the deviant sounds (Lachmund 1994, 1999; Rice and Coltart 2006; Alberts 2000, 2003). Whereas monitory listening refers to determining "*whether* something is wrong," diagnostic listening is intended to reveal "*what* is wrong" (Pinch and Bijsterveld 2012: 14).

Such modes of listening also structured the social practice of "listening while driving" in the early twentieth-century automobile (Krebs 2012a: 83). Automobilists were not only supposed to learn the technique of monitory listening, but also of diagnostic listening. "For the detection of technical failure," a handbook stressed, "noise is crucial" (Hessler 1926: 216). Diagnostic listening, however, was much harder to achieve than monitory listening: "This skill, to make the correct diagnosis out of a knocking sound, requires tremendous experience and exact knowledge of the type of engine construction" (Hacker 1932: 83). Maintenance almanacs for motorists provided help, though: they offered a systematic overview of possible malfunctions, their symptoms, and ways of repairing these problems. Oskar Hacker's manual (1932) even categorized malfunctions by the senses with help of which they could be detected: seeing, hearing, smelling, and feeling. He needed no less than twenty-nine pages to list a wide array of audible failures, an approach also adopted in automotive journals. Other manuals ordered the maintenance section by car components or engine functions, but did as much to explain the sounds that would be audible when problems might occur (Schmal 1912; Küster 1919). Implicitly, handbooks began to codify car sounds by transforming auditory experience into communicable signs and meanings, if only with limited success, as we will see later. Jens Lachmund has described similar problems for physicians when they started to listen to a patient's body with the help of the stethoscope. Hearing things was one thing, yet ordering sounds, relating them to pathologies, and explaining the character of the sounds to others was quite another (Lachmund 1999: 420; see also Rice 2012).

All of the handbooks emphasized that a systematic inquiry was crucial. To assist the motorist in achieving this level of rigor the manuals contained tables, lists, and fault trees. They could be used to perform a differential diagnosis to narrow down the list of potential problems to a single condition and provide a basis for a hypothesis of what was troubling the "patient" (Hacker 1932: 18). Subsequently, the motorists had to unravel the links between the symptoms, the components that might be "infected," or the specific driving conditions under which the malfunction was most noticeable—just as a physician would take a patient's medical history. "If the physician cannot make his diagnosis by the appearance of the patient, he will take his stethoscope and listen to the patient's body. This is how you ought to proceed with your car engine as well" (Hessler 1926: 216).

In the illustration essay at the end of the previous chapter, we discussed an illustration of a doctor-like person listening to a car engine with a stethoscope (see figure 2.17). Here the reference needed to reassure potential owners that in a car with "silent" Timken Axles "[n]ot even a stethoscope could catch the pulse of this artery of power transmission" (advertisement, Timken Axles 1932). Manuals, however, advised drivers to use instruments to listen to specific spots for diagnostic reasons. Motorists could, for instance, employ a screwdriver or a long metal pipe as simple ear trumpet (Hacker 1932: 81). A driving manual issued by the German Association of Motorists claimed that, with a stethoscope and a trained ear, one could locate a single dry-running bearing (Dietl 1931: 324). Listening devices, such as the Auto-Doktor, were advertised in automotive journals for a wide audience (Anonymous 1929a). The implication was that motorists could develop the same level of sonic skills—with the proper use of auditory instruments—as physicians.

Dutch-language manuals for motorists expressed equal trust in what drivers were able to learn in the auditory realm. Aberrations in the normal functioning of cars, Dutch engineers stressed, could be heard long before they could be seen (Snook 1925: 115; Geluidstichting 1936: 77). This prompted authors of driver's handbooks to include advice on how motorists should listen to their car. A "sneezing" car was saying something about an obstruction in the carburetor's mixing chamber. A "trained ear" would hear that this sneezing did not start all at once, but gradually announced itself through a change in the pitch and rhythm of the engine's sound. "Pinging" and "knocking," the more complex aspects of the car's "grammar," told of even worse problems (Meyer [1930?]: 19, 20, 22, 24; see also Martini 1932: 81, 164, 203, 222, 371–72; Ganzevoort 1955: 99–100; Joppe 1958: 119–20; de Graaf and de Rétrécy 1961: 109–13). As the publication dates of these manuals reveal, this happened as late as the early 1960s. Already in 1947, however, a handbook noted that one needed much professional expertise and routine to diagnose car problems from the "tone of the sounds" (Brand 1947: 340). And a German manual added to most of the descriptions of ominous sounds that the motorist should see a mechanic as soon as such distressing noises popped up (Dillenburger 1957: 301–3). This addition, as we will explain below, signaled shifts in the professional position of mechanics in Germany.

In 1920s Germany, handbooks still highly valued drivers' diagnostic skills and assumed that they were eager to acquire such expertise. Authors of manuals expected this because repairing a car oneself was cheaper than paying a mechanic and, if done in time, useful in avoiding more serious problems. Moreover, owners were legally obliged to monitor their cars to avoid accidents that might occur due to mechanical problems, even if they employed a chauffeur. Employers thus needed technical knowledge to supervise their chauffeurs, in particular if they doubted their trustworthiness (Anonymous 1926). Reading the "Letter Box" published in the *Allgemeine Automobil-Zeitung* helped motorists to acquire such expertise, as these letters were often about technical issues. Between 1928, the start of this feature, and the outbreak of World War II, the journal received some fifteen thousand letters from readers, a selection of which ended up in the "Letter Box."

Some of these letters disclosed information about motorists' listening practices. Owners indeed carefully monitored how their machines ran: they gave detailed accounts of the specific conditions under which a suspicious noise occurred. And they used a wide range of words to capture the audible idiosyncrasies of their automobiles. Their cars were sobbing, whining, rumbling, stuttering, hammering, knocking, singing, howling, growling, ticking, hissing, droning, or even chirping like crickets (Briefkasten 1928c, 1930). Other motorists described the frequency and pitch of the noises (Briefkasten 1929a). Still others clarified which tests and repairs they had already carried out themselves before asking advice, or stressed how knowledgeable they were. "I have been a self-driver for twenty-two years now, and I know a lot about engine designs, but this time I am helpless," one driver grumbled (Briefkasten 1928d).

Often the editors had difficulty making sense of the accounts because motorists, despite the attempts to codify sounds in handbooks and journals, shared no standardized vocabulary to describe their auditory experiences. Knocking, for example, might indicate spontaneous ignitions or worn-out piston bearings. In difficult cases the editors gave rather general suggestions or explained how to narrow down the range of possible faults (Briefkasten 1929b). At times the communication failed completely. One reader wrote that a "hot noise" made no sense to him, while the editors stressed their know-how by claiming that experienced mechanics needed more than ten years to develop a "trained ear" (Briefkasten 1938, 1933). Nonetheless, the advisers always treated the readers as technically competent and capable of doing major repairs (Briefkasten 1928a).

It is clear, however, that their intended translation of embodied expertise was not invariably successful. We will return to this tacit dimension of knowledge below. Here it is important to repeat that monitory and diagnostic listening were seen as skills a competent driver mastered. Such skills became a mark of distinction for motorists in the 1920s. As Andrea Wetterauer has claimed for Germany in these years, repair expertise matched upper-middle-class social standing, and belonged to the cultural techniques of the bourgeois: as long as driving was relatively exclusive and expensive, the associated technical expertise was a sign of social importance (Wetterauer 2007, 155–66). Similarly, the ability to listen to and understand one's car expressed, in terms of Marcel Mauss's work (1936), the *techniques du corps* of a distinguished driver. Such techniques of the body implied, in Pierre Bourdieu's version of the idea, a "*socially informed body*, with its tastes and distastes, its compulsions and repulsions, with . . . all its *senses*, that is to say, not only the traditional five senses . . . but also the sense of necessity and the sense of duty, the sense of direction and the sense of reality" (Bourdieu 1977: 124; see also Bourdieu 1990, 1999). What's more, motorists' diagnostic capabilities could make up for their potential distrust of chauffeurs and, as we are soon to learn, of car mechanics.

LISTENING TO THE AUTOMOBILE: THE MECHANIC'S EAR

Despite great expectations about German car ownership at the end of World War I, mass motorization did not take off in Germany during the interwar period (Ruppel 1927). Still, the German car-repair business displayed a strong growth, with some 20,000 repair shops in business in 1929 (Reparatur-Werkstatt 1929: 1–2). This figure included a huge number of workshops run by blacksmiths, tinsmiths, and fitters, who repaired cars on the side, "ad-hoc mechanics," as automotive historian Kevin Borg has called them for the United States (2007: 31–52). In contrast to other trades in Germany at that time, the work by auto mechanics had not yet been legally regulated. It was thus legitimate for virtually anybody, including chauffeurs, retired army drivers, and the variety of smiths just mentioned, to repair cars.

Journals for those interested in car mechanics frequently ran articles on sound and listening. A special section in *Auto-Technik* entitled "For the Repair Shop" gave advice on how to get rid of minor noises, such as the rattling of the brake linkage. More in-depth articles explained the technological background of the new noiseless chain drives or the complex phenomenon of engine knock (Ostwald 1921, 1922)—two issues already introduced in the previous chapter. The best instrument for determining the antiknock quality of fuels, one author claimed, was a well-trained ear—nothing as sensitive to differences in frequency (Enoch 1928). Focused listening was considered highly important for the diagnostic potential of car mechanics (M. S. 1928). As an anonymous author put it, "For diagnosing engine sounds a very fine sense of hearing indeed is required, especially when the sounds are very faint and if several sounds from different sources have to be distinguished simultaneously, which is often the case" (Anonymous 1932: 81).

The same author claimed that a lack of experience could not be compensated for by using a listening device. If mechanics were not used to working with "a wooden rod or a stethoscope"—which had to be positioned outside of the engine—they would "easily be misled by the effect of resonance" (Anonymous 1932: 81). Other authors were much more optimistic. Such devices could certainly assist the mechanics, especially when they had to concentrate on specific spots in the machinery. *Auto-Anzeiger*, a trade journal not distributed to ordinary motorists, presented a stethoscope with two sensors: the Tektoskop and the Tektophon. This type of construction enabled the mechanic to examine two engine spots at once, thus allowing him to compare two sounds in detail (Anonymous 1929b). Another tool, the Meccano-Stethoskop, was labeled as "the ideal troubleshooter." It would help the mechanic to "clearly observe the processes inside the engine" and to save time as disassembly of the engine was not required when diagnosing problems (Anonymous 1930d). This claim strongly appealed to workshops, as customers expected a prompt and correct diagnosis.

Because making a diagnosis through sound was so hard, however, *Auto-Technik* warned the mechanics among its readers to give only non-binding estimates of the trouble (Walkenhorst 1926). Its warning testified again to the complexities of diagnostic listening, which required both sound verbalization and "sound mapping." The notion of sound mapping comes from the field of psychoacoustics. It refers to the connection between sounds and the information they represent (Fricke 2009: 55–56) and is a useful metaphor for grasping diagnostic listening as associating particular car sounds with specific malfunctions. The concept of mapping helps to explain why stethoscopes and other hearing devices alone were of little help. It was not sufficient to simply amplify sounds—the users of the tools also had to assign meaning to them. But how did car mechanics learn to do so?

The literature mechanics had available—as had been the case for motorists—agreed on the importance of theoretical knowledge, practical experience, and a systematic approach as preconditions for proper listening (Anonymous 1928; Winkler 1928; Fischer 1927). Authors kept struggling with translating the audible signs of malfunction into words, however. The phenomenon of engine knock is, again, a good example. One author described engine knock as a hammering sound, while the only sound really *knocking* was that of a malfunctioning piston bearing (Anonymous 1919). Another author distinguished no less than seven types of knocking, including "metallic knocking," "high knocking," "damped knocking," and a "muffled clang" (Anonymous 1932). Apparently, a standardized set of subtly differing terms to express the audible characteristics of car problems was not available. So how then could the mechanic know what these sounds referred to? Was written advice really helpful? Eugen Mayer-Sidd, contributor to the car mechanics journal *Krafthand*, had his doubts about the functionality of verbalizing car repair knowledge, even though he had often given written advice himself: "It is exceptionally difficult to give someone else a detailed and graspable description of a technical work or method that enables him to do it himself later on" (Mayer-Sidd 1931b).

The unachieved codification of car sounds and the difficulties of giving written advice underscore the relevance of the tacit dimension of car repair work. Michael Polanyi identified such "tacit knowledge" as the nonexplicit aspects of knowledge that are hard to transfer from one person to another—the kind of knowledge that novices can only acquire by observing experienced people (1958: 69–245). In a similar way, sociologist of science Harry Collins speaks of the "unconscious emulation" of knowledge that is not cognizable in verbalized terms (2001: 72). It is not surprising, then, that extensive training in car repair was considered of increasing significance for transferring tacit repair knowledge, as our next section will show. Tacit knowledge, however, can also easily be distrusted. Let us explain how the tacit knowledge of the car mechanics required legal protection in order to survive as their acknowledged expertise.

THE REPAIR CRISIS AND THE PROFESSIONALIZATION OF THE CAR MECHANIC

At the end of 1926, the German Chambers of Industry and Commerce published an assessment of the field of auto mechanics. It was quite critical, mentioning recurrent complaints about excessive prices for spare parts and repair work. Car dealers denied these shortcomings (Anonymous 1926). Yet the *Allgemeine Automobil-Zeitung* published several letters by readers articulating a rising distrust and dissatisfaction among owners (Briefkasten 1928a, 1928b).

In 1928 *Auto-Technik* published an editorial under the heading "The Great Repair Misery." Its author complained about the trustworthiness of mechanics employed by dealers and manufacturers, as well as mechanics working in independent repair shops. During the warranty period, for example, the significance of audible technical problems was often played down if not denied by the manufacturer's mechanics: "They try to persuade the customer that an abnormal sound, which indicates an upcoming problem, is of no significance—'this does not mean anything,' 'this is just an imperfection'" (Loewe 1928: 11). In some cases motorists discovered only after the "repair" that a clearly audible problem had simply not been tackled. The author of the editorial interpreted this as signaling a lack of expertise among the mechanics. The repair misery, he argued, posed a serious threat to the automobile system because dissatisfied motorists might abandon their cars. An editorial in the *Allgemeine Automobil-Zeitung*, then, asked readers to suggest a way out of this "repair chaos" (Anonymous 1931b).

Advocates for the repair business, however, did not seek an answer to the repair crisis among motorists, but pointed to the unregulated access to the trade: "The blacksmith, the bicycle or sewing machine mechanic, the fitter, they all have to learn their trade for four years, but anybody who learned to handle a file and followed a six-week course is able to repair the complicated and valuable engine of a car flawlessly?!" (Testor 1931). To put things right, they proposed the establishment of an independent car mechanics trade together with mandatory membership in a guild (Anonymous 1931a). In Germany, laws governing trade and industry restricted the right to enter a recognized trade or guild to those who had done a three- to four-year formal apprenticeship and had passed a final exam for acquiring a journeyman's certificate (*Gesellenbrief*). After three to five years of additional experience, journeymen achieved the right to take a second exam to obtain a master craftsman's certificate (*Meisterbrief*). Only with this second certificate did they have the right to train apprentices (Greinert 1994). German craftsmen thus cultivated a preindustrial mentality, a habitus grounded in a long and painstaking apprenticeship, in which a sense of "master craftsman's honor" and the ideology of "high-quality workmanship" served as guiding values (Holtwick 1999; Sennett 2008).

It was the trades' regulated position *and* the associated "symbolic capital," to use one of Bourdieu's concepts (1989) again, that attracted the representatives of the repair workshops. Creating a car mechanics trade guild would structure the relationships between members and nonmembers and grant the mechanics formal jurisdiction over a particular field of expertise, an important phase in processes of professionalization, as Andrew Abbott (1988) has shown for medical and other professions. In addition, the new-style mechanic would develop a strict sense of responsibility during his apprenticeship under the guidance of a master craftsman. Just like a physician, the future mechanic would embody both expertise and commitment to good practice. It was thus no coincidence that the car mechanic became increasingly represented as a car doctor. The links between motorists' repair knowledge and medical expertise had been somewhat fuzzy, but now the car mechanic was simply dressed up like a physician, at least in the illustrations in figures 3.1 and 3.2.

This professionalization, so ran the idea, would restore customers' trust in the repair business. The repair lobby did not immediately have its demands accepted, however. Only in 1934 did the National Socialists—just in power—introduce legislation that made guilds obligatory for *all* trades, thus abolishing the existence of unrecognized trades. A year later, they established the master craftsman's certificate as a precondition for starting a workshop, whereas this had only been a requirement for training apprentices previously (Winkler 1972: 184–85; Saldern 1979). The first to profit from this novel legislative framework were the car mechanics, a result that meshed well with the National Socialists' plans for mass motorization of Germany (Zeller 2010 [2006]). The connections between the new regulations and the long-standing traditions of the guilds resolved the crisis of confidence between motorists and car mechanics, because the symbolic capital of

Figure 3.1
Illustration from the *Allgemeine Automobil-Zeitung* (AAZ), "Letter Box" section, depicting the image of car mechanics as doctors (1932)
Source: Jonny 1932.

the honorable certificates equipped mechanics with unquestionable expertise and trustworthiness, independent of their individual skills.

How important this situation was for the status of the mechanic becomes clear when we compare it with the North American state of affairs. In the United States, the trustworthiness of the car mechanic remained contested due to a continuing lack of regulation concerning access to the trade. To solve this issue, mechanics sought to delegate diagnostic authority to instruments and measuring devices that visually displayed the results—but failed. The introduction of a flat-rate system did not help either because mechanics worked hastily, doing a shoddy job under the pressure of standardized fees (McIntyre 2000: 292). As a result, American car mechanics have suffered from an endless crisis of confidence (Borg 2007).

In contrast, the expertise of German auto mechanics was guaranteed through their obligatory four-year apprenticeship. During these years

Figure 3.2
Mahle advertisement, detail (1940)
Source: Advertisement, Mahle 1940.
Courtesy: MAHLE International GmbH.

the future mechanic worked in close contact with a master craftsman or journeyman and learned through observation and imitation. To ensure the quality of this process, a master craftsman was not allowed to train more than two apprentices at the same time. Starting with simple jobs such as checking the tire pressure, the apprentices gradually acquired more responsibility, doing increasingly complicated maintenance work and learning metalworking skills such as filing, drilling, milling, turning, and welding (Kümmet 1941; Zogbaum 1937). The training manuals for the novices did not explicitly mention specific sonic skills such as diagnostic listening. Apparently, such skills were now considered part and parcel of the everyday learning-by-doing training. This is indeed, as sociologist Douglas Harper has shown, the way in which car repair knowledge is informally passed on from one person to the other (1987: 24–31). Yet what happened to the listening skills of the ordinary driver?

"A MARRIAGE ON ITS LAST LEGS": THE DIFFERENTIATION OF LISTENING PRACTICES

"Hands Off" was the telling title of an article published in the *Allgemeine Automobil-Zeitung* of May 1933. Its author urgently requested readers not to repair their cars themselves: they simply lacked the necessary abilities. They should also stop bothering about little noises that were just a nuisance, and should only bring their car to a specialized repair shop if they really heard a "threatening noise" (Anonymous 1933). With the rising number of motorists, another automotive journalist claimed, the relative number of "knowledgeable motorists" had declined. He assumed that the high level of technical quality reached in the automotive industry and the increasing reliability of cars explained this change, but he also considered most people just "terribly clumsy" (Rdl. 1936: 18, 19).

This critical tone concerning the technical expertise of drivers of course also expressed the trend toward professionalization of repair work discussed in the previous section. The same was true for three other remarkable developments. First, the journals for motorists displayed a rhetorical shift from talking about *repairing* to writing about *tinkering*. Instead of advising readers how to repair cars, authors tried to make them enthusiastic about tinkering with their cars, as a new pastime (Rdl. 1938; see also Franz 2005). Second, a new official master craftsman's handbook proposed a ground plan for garages that rigorously separated motorists and mechanics: the showroom and a waiting room for customers served as the garage's front office, while the workshop proper was its back office, for mechanics only (Kümmet 1939: 250). Third, the publication of cartoons and short stories about naive drivers additionally had to demarcate genuine expertise from amateurish messing about with cars, and underlined the mechanics' new self-awareness as professionals. Stories had both sloppy and neurotic drivers as their antiheroes (Jonny 1937; Windecker 1937). Whether owners were lazy or overactive, the message was that they should keep their hands off their cars' technology.

Other stories articulated the struggle for demarcating *listening expertise* from *listening ignorance*. "As you know," one author addressed his colleagues, "there are so-called noise fanatics who can drive a busy master craftsman crazy with their accounts of noises, sometimes real and sometimes imagined, they have heard" (Anonymous 1938). Another article characterized this type of driver as someone who "is often bothered by noises that exist only in his imagination. The fact that he is always sure where the noise comes from does not make him more likable because he is usually wrong, thus leading the craftsman down the wrong road." With the topos of the "noise fanatic," mechanics reclaimed the practice of diagnostic listening as their exclusive domain: "When looking for a noise source, never ever let yourself be influenced by the customer" (Anonymous 1939). By denouncing such drivers as overanxious and unknowing, mechanics deprived listening motorists of their expertise.

It was not that motorists had to stop listening to their cars altogether. Even though the "layman" was "often anxious about harmless noises," it was "still better for him to consult an expert in vain than to disregard noises until the engine has a serious problem" (Anonymous 1936). The driver, in other words, should be neither a "noise fanatic" nor a "noise stoic." A noise stoic was the type of driver who would not bother to notice even when the car's chassis played "a free concert" with the engine. Because of his "tin ear," such a "symphony does not disturb him; he will not do anything until his heap breaks down. We find these people just as disagreeable" (Anonymous 1939). Clearly, the driver should not abstain from monitory listening. Diagnostic listening, however, had to be left to the mechanics. Drivers ought to be realistic about prices and the time repairs took (Dill 1936) and should, first of all, trust their mechanics: "If your car is more important to you than your rhetorical exercises, let the master craftsman do the job" (W. 1938).

Over time, even monitory listening lost some of its significance in connection to everyday driving. In the early sixties, Alexander Spoerl, German author of a concise manual for motorists that also appeared in Dutch, used an intriguing analogy to describe the new auditory relations between the driver and his car. He compared the process in which the driver started feeling annoyed about the acoustic peculiarities of his car with a marriage on its last legs.

> [In a flirt] with a girl, or someone else's wife, all is fun and pleasure. But when it comes to our own wife all sorts of trifles start to make us nervous, even to the extent that we begin to scheme a murder or suicide.... This is how it goes with cars. The rattling taxi is not ours... and does not disturb us in the least. However, once we... are married to our own vehicle, there is always something that starts annoying... us. (Spoerl 1963a: 198)

Such annoying sounds, Spoerl continued, could be a ticking valve, a stammering carburetor, a pinging engine, a creaking gearbox, or a "suspicious whistle." Your car "starts falling to pieces" and other motorists start overtaking you, some of whom are just "out of the cradle" (Spoerl 1963a: 198–201).

This author clearly thought poorly of marriage—which was a common topic in the automobile literature of the 1950s and 1960s. More important, however, is the observation that Spoerl no longer considered the driver's listening to the car's noises as functional—now such noises were just a nuisance. He may have been exceptionally outspoken, as driver manuals published in the Netherlands in the 1960s and early 1970s still occasionally referred to the sounds of malfunctioning, notably pinging and knocking. Yet distinct chapters on how to listen to one's car or long lists of sounds disappeared (Berk 1961; Anonymous [1969]; Hinlopen 1971; Olyslager 1971). Nonprofessionals, as one of these manuals stressed, usually lacked the experience to "diagnose by ear." And it was hard to explain "on paper" why "that one 'thudding sound' was caused by a problem with the ignition, while another 'thudding sound' was brought about by an obstructed acceleration jet" (Olyslager 1971: 307). Spoerl's reference to young cars might even indicate that excessive noise should be interpreted not as a sign of needed repairs, but as an incitement to find a *new* car that ran properly, or, as he also suggested, to have two cars and alternate their use (Spoerl 1963a: 202). As transport historians Gijs Mom and Rudolf Filarski have shown for the Netherlands, the practice of preventive maintenance—the garage offering to replace a particular component before it wore out—started in the 1930s and was common practice by the 1950s (Mom and Filarski 2008). This may have been another context relevant for the disappearance of sound mapping from the motorists' manuals.

Trade journals for car mechanics continued to publish articles on how to deal with audible malfunctions (see, for instance, Anonymous 1955). Drivers, however, were no longer supposed to listen to the sound of cars in the focused manner that had become the jurisdiction of mechanics. Early interior noise control, as we have seen in the previous chapter, had tamed some of the sounds that could bother the driver, and the closed car body had encapsulated the motorist, now separated from the outside environment. In addition, the mechanic had started to tell the driver not to listen more carefully than was necessary to distinguish fair and foul. All this happened in the first three decades of the twentieth century, and together these developments freed the ears and attention span of drivers for listening to new sounds. One source of such sounds was car radio. From the 1920s onward, radio's voices and music started to fill the space in and around the car. Yet, as the next section will clarify, the naturalization of listening to radio while *driving* a car required a special effort from the industries and experts involved.

DECENT RADIO RECEPTION—AS LONG AS THE CAR IS NOT MOVING

> The newest thrill of this thrilling age is a radio set installed in your automobile—and the radio set for automobiles is Transitone. It is recognized by motor car manufacturers and public alike as the greatest contribution to motoring pleasure in recent years. . . . Find out about it. Learn the thrill of having music with your mileage. (Transitone 1930, reprinted in Matteson 1987: 75)

This 1930 triple "thrilling" ad for Transitone, one of the first built-in car radios offered in the United States by Automobile Radio Corporation, expresses nothing but confidence in the quality of car radio. Transitone's trademark, a music note with wings attached, reminds us of the dream of the smoothly flying car discussed in the previous chapter. Car radio, however, did not mix music with mileage as easily as the ad suggested.

Transitone was introduced on the American market in 1928.[2] Over the next few years, other makes followed suit, such as Motorola and Crosley, and, in Europe, Blaupunkt and Philips. One might assume that the moment of the radio's integration in cars was linked to the introduction of the car's enclosed body, which also occurred in the 1920s, as we have seen. Without a closed body, the sounds from the surroundings and wind noise would drown out the radio. Yet the early ads for radio were not exclusively geared to radio listening in an *enclosed*, *moving* car, but also to listening in a *parked*, *open* car, and not just by those inside the car but also by those outside.

Little research has been done so far on the origins of car radio. We do know, however, that American radio amateurs started experimenting with car radio years before the introduction of Transitone.[3] In the early 1920s, motorists took portable radio receivers along in their cars, and in 1922 Chevrolet had a sedan equipped with a relatively expensive radio (Klawitter et al. 2005; McDonald 2008a, Matteson 1987: 33). To understand this enthusiasm for bringing radio receivers along, we need a basic grasp of the role of radio in the daily life of Americans at that time. In the 1920s, the public interest in radio was huge. As of 1930, as many as 40 percent of the households in the United States owned a radio. Ten years later this figure had reached over 80 percent (Butsch 2000: 176, 205). Moreover, by

2. Philips Company Archives (PCA), Eindhoven, the Netherlands, File 811.215, 1928–1932, handwritten note with a reference to Radio Express, November 30, 1928.
3. PCA, File 811.215, 1928-1932, Letter L. P. Graner to P. Staal, May 21, 1929.

the early 1930s it had become normal to combine radio with other activities (Douglas 1999: 84). Those who went out camping or picnicking by car might thus well think of taking along a radio.[4] And most likely, radio manufacturers were looking for new markets. Mobile radios were not only introduced in passenger cars, but in police cars and fire engines, in taxis, busses, trains, airplanes, on boats, and even on bicycles (Schiffer 1991: 113–15; Weber 2008: 144).

In the United States car radio grew popular fairly quickly, even if the first years were difficult for the manufacturers because of the Depression. Still, by 1941 30 percent of all automobiles had a radio, while 50 percent of all *new* cars sold that same year were equipped with a radio (Butsch 2000: 206). In Europe its introduction was much slower. In 1971, for instance, when almost 95 percent of the US cars came with a radio (Fesneau 2009: 314), just over half of the cars in West Germany had one (Weber 2008: 142). A German automobile journalist explained the early success of car radio in the United States with reference to the country's dense network of shortwave radio broadcasting stations (Lutz 1939: 1078).

In ad materials geared to the American market, car radio was initially associated with romance. Although Philips introduced its first car radio in 1934, a 1952 commercial—the oldest one we were able to trace—still eagerly taps the romantic tradition. If this commercial presents an American setting, we hear Dutch voices. We first see that an open, radio-equipped car is being overtaken by a closed car. Upon arriving at some mansion, this car's driver honks loudly. A lady shows up on the mansion's front balcony, but she does not respond. A little later the open car arrives at the same location and its driver "honks" by raising the volume of his car radio. This time the lady responds right away by waving, and a moment later she gets into the car. Given this plot, the commercial's title hardly comes as a surprise: *The Secrets of his Power*.[5]

A substantial share of the Philips prewar and immediate postwar ad materials in fact conveys that a car radio, commonly shown in an open automobile or one with an open door, attracts the attention of an admiring woman or audience (figures 3.3–3.5).[6] Surely, Philips thus sought to hook up with what the company considered to be the preoccupations of

4. PCA, File 811.215, 1933 ff., C. Tindal, "Auto en Radio," 1934, *Het Motorrijwiel en de Populaire Auto* (February 16): 196–98, at 196.

5. PCA, "The Secrets of His Power," video, 1952.

6. PCA, File 812.215, photos of ad materials, 1934–1960, illustrations 1934, 1939, 1949, 1953, and PCA, File 811.215, Product Documentatie Autoradio's, "Philips Auto Radio Tips,"1936 (5, May–June), illustration following page 1.

Figure 3.3
Philips advertisement (1936)
Source: Philips Company Archives Eindhoven, File 811.215, "Product Documentatie Autoradio's." Philips Auto Radio "Tips," no. 5, May–June 1936, 10th page (no page numbering).
Courtesy: Philips Company Archives.

young men, many of whom were among what innovation scholar Everett M. Rogers has coined the "early adopters" of new technology (Rogers 2003 [1962]: 283). But from sources in the Philips company archives it can be deduced that Philips' commercial strategy was inspired by other issues as well. In those years Philips worked with several foreign correspondents, informants who would visit one convention after the other, take note of new technologies introduced by competitors, and report on what was written about them in local newspapers. From 1929, the Philips correspondent in the United States, consulting engineer L.P. Graner, reported on the first built-in car radios.

In Graner's view, these car radios were expensive and of poor quality. With a copper or aluminum antenna on the roof and a convenient placement of batteries and loudspeaker, the reception was quite decent—as long as the car was not moving. Yet when driving, the amplifier tubes easily failed, while reception was poor on account of interference caused by the ignition system's electromagnetic waves. Similarly problematic was that the engine often produced a higher level of sound than the radio. The result, according to one contemporary, was "a hellish noise of distorted

Figure 3.4
Philips advertisement (1939)
Source: Philips Company Archives Eindhoven, File 812.215, Advertisement material 1934–1960.
Courtesy: Philips Company Archives.

sounds, accompanied by all sorts of parasitizing sounds."[7] Ironically, this was a far cry from the earlier expressed engineers' hope, mentioned in the previous chapter, that the radio's sound would mask the car engine's noise!

The introduction of screen grid fuses made amplifiers more solid, but countering interference proved harder. Engineers experimented with covering the radio receiver, shielding the cabling, and adding resistances and condensers to the ignition system, but this was said to weaken the energy transmitted by the sparkplugs and thus the engine's performance (Anonymous 1948: 226; McDonald 2008b: 6).[8] In the solutions to the

7. PCA, File 811.215, 1933 ff., W. Vogt, "Radio in een Auto," *De Auto* [February 15, 1934]: 253.

8. PCA, File 811.215, 1933 ff., Th.P. Tromp, Rapport No. 1 Reis Amerika, February 16, 1934. See also PCA, File 811.215, 1928–1932, "Installs Radio in Car With Good Results," 1929, *The New York Sun Radio Section* (December 7); Letter L. P. Graner to P. Staal, May 21, 1929; Letter L. P. Graner to N.V. Philips Radio, Eindhoven, December

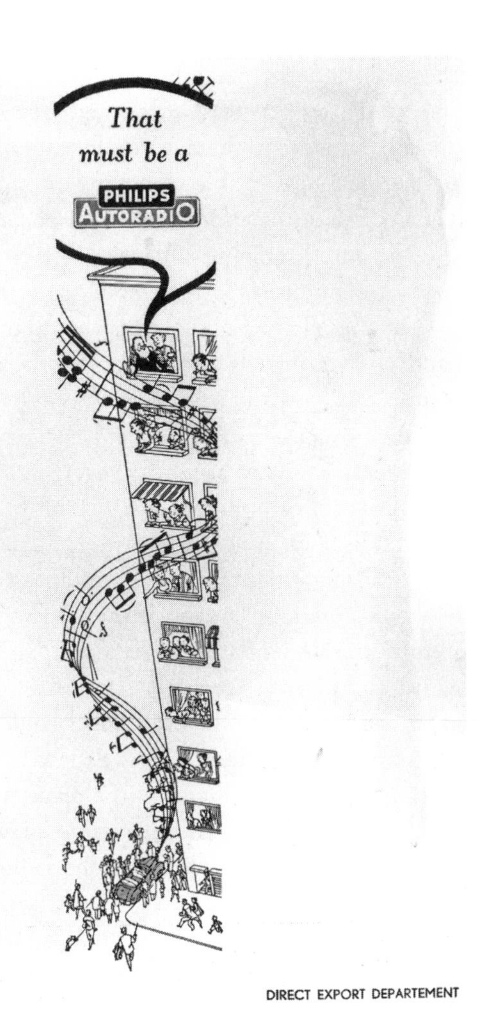

Figure 3.5
Philips advertisement (1953)
Source: Philips Company Archives Eindhoven, File 812.215, Advertisement material 1934–1960.
Courtesy: Philips Company Archives.

problem of interference, then, the interests of carmakers clashed with those of radio manufacturers. The same was true regarding the size of the batteries. Because radios used a lot of power, car batteries went dead quite soon. Car manufacturers, however, did not want to add heavy batteries to power a radio because that would degrade car performance. Philips

20, 1929; Arthur H. Lynch [1930], "Radio takes the open road"; Stuart C. Mahanay, 1930, "How to Install an Auto-Radio Receiver," *Radio News* (February): 722–73; "Jedes Auto ein Radioapparat," 1930, *Radiowelt* (March 9–15); "Putting Radio Sets in Automobiles," 1930, *The New York Sun Radio Section* (April 19); W. H. Smits, 1932, "Radio-ontvangtoestellen in automobielen," *Polytechnisch Weekblad* 26 (17).

resolved these tensions by making a design whereby the radio was connected to the car battery instead of having its own batteries. In addition, Philips made adjustments to the engine superfluous by rendering the radio equipment itself less susceptible to distortion.[9] Such solutions were not unique. In fact, much of the effort of both European and US-based firms in the 1930s[10] went into problems of energy supply and distortion. In addition, the position of antennas (often embedded in the roof, yet externalized with the rise of the all-steel body), superheterodyne reception, dynamic speakers, and push-button automatic tuning were supposed to improve the radio's sonic qualities and ease of operation (Matteson 1987: 41–220; Duckeck 1973).

An issue at least as substantial, however, was the social unrest that erupted over the use of car radio. Soon after this technology's introduction some traffic officials began to express concerns about how it would distract motorists and cause accidents, notably in areas with dense traffic. Such areas were a noisy madhouse even without car radio. Various American states therefore considered banning car radios, either within city limits or altogether. The city of New York was especially worried about the use of radios in taxis. By paying with a coin, clients could request taxi drivers to turn on their radio. But after an elderly pedestrian was hit by a taxi, this practice became controversial because the driver was allegedly distracted by his radio. The state of Massachusetts came up with another solution to this problem: one was allowed to operate the radio only when the car was parked, not when driving.[11]

Initially, drivers seemed to share such concerns. Historian David Goodman cites a 1934 survey by the Automobile Club of New York that showed that "56 percent of the respondents regarded car radios as a dangerous distraction" (2010: 37). Yet the manufacturing industries successfully lobbied against regulation, and a widespread car radio ban never materialized. And when in 1939 Edward Allen Suchman of the Princeton Radio Research Project published a study on the use of radio by New York cab

9. PCA, File 811.215, 1933 ff., Th.P. Tromp, Rapport No. 1 Reis Amerika, February 16, 1934, p. 3.

10. By 1935, the United States had over eighty such firms (Matteson 1987: 147).

11. PCA, File 811.215, 1928–1932, Letter L. P. Graner to N.V. Philips Radio, Eindhoven, December 20, 1929; "Putting Radio Sets in Automobiles," 1930, *The New York Sun Radio Section* (April 19); O. Bruno, 1930, "Automobil und Radio," *Radiowoche* (19, May 7): 3; Letter L. P. Graner to N.V. Philips Radio, Eindhoven, June 13, 1930. See also PCA, File 811.215, 1933 ff., "Autoradio gevaarlijk op den weg?" 1934, *Het Volk* (September 19); File 811.215, Product Documentatie Autoradio's, "Vijftig jaar Philips autoradio," 1984, *Philips Koerier* (April 12): 5. See also Suchman (1939).

drivers and its effect on rates of accidents, his conclusion was that "radio has little effect upon the safety of taxicab operation" (Suchman 1939: 153).

However, when Suchman interviewed drivers in general and taxi drivers about car radio, they gave arguments both for and against. They reported that "exciting broadcasts" distracted the driver, that the sound of radio masked warning signals like train whistles, and that "the manual operation required" for radio diverted "the eyes from the road." Among the arguments in favor of car radio were that it induced "slower driving" in order to enjoy the program, that it kept drivers "alert," and that it reduced "disturbing conversation" and "backseat driving" by other occupants because passengers were now "entertained and therefore more quiet" (Suchman 1939: 154). Moreover, when Suchman found that drivers of cars without radio were more negative about it than those with car radio, Suchman suggested that the effect of car radio might be influenced by a form of internal self-correction. If radio-sensitive car owners did not install radio in their cars and non-radio-sensitive did, this would also "help account for the lack of accidents, since those people having radios in their cars are radio-non-sensitive to begin with" (Suchman 1939: 155).

For several years it was uncertain how this controversy would be resolved. Against this backdrop, it is understandable that car radio in a *nonmoving*, open car received attention from advertising agents, at least until the early 1950s. In addition, in its early marketing of car radio in Europe, Philips construed both literally and figuratively a space of its own by stressing that car radio was meant to be listened to in the countryside, not in busy urban traffic. In the city, according to Philips, motorists indeed needed to pay attention to other traffic.[12] A 1931 test drive in the vicinity of the company's factories in Eindhoven had shown the Philips employees that the Transitone's reception was quite poor, notably in narrow streets with high buildings. This was not a problem outside the city, where listening to the radio was "a pleasant sensation" and not dangerous.[13] In the countryside radio encouraged the driver's sustained attention. Those who had their radio on would be driving more slowly and hence more safely than someone who was not in contact with the outside world.[14]

12. PCA, File 811.215, Product Documentatie Autoradio's, "Philips Auto Radio Tips" 1936 (5, May–June): 2; File 811.215, 1933 ff., "Autoradio gevaarlijk op den weg?" 1934, *Het Volk* (September 19).

13. PCA, File 811.215, 1928–1932, Radio in auto, Demonstratie op Dinsdag 5 mei, door den Heer Keurborst, vertegenwoordiger der Crysler Corporation, Verslag Van Thijn, May 6, 1931.

14. PCA, File 811.215, Product Documentatie Autoradio's, "Philips Auto Radio Tips," 1936 (5, May–June): 3.

Philips salesmen received extensive information on how to pitch the product in this way.[15] "A Philips car radio makes the drive more pleasant and keeps the mind alert," as a 1937 brochure put it.[16] Radio was an "entertaining travel companion on lonesome journeys" that helped prevent the driver "from occasional lapses while driving" and turned "driving at night into a pleasure."[17] Those who drove a car with a radio would never be lonely again. One of the Philips ads shows a woman who at night is sitting behind the wheel alone. The darkness is eerie, yet the woman imagines a romantic violinist as companion (figure 3.6).[18] Such companionship would drastically shorten the distance to be covered, it was suggested, and replace "monotony with polyphony."[19] Philips also stressed that through car radio "all of Europe" came "within easy reach," since the Philips sets, in contrast to the American sets, received not only shortwave broadcasts, but also longwave, a frequency used commonly in Europe.[20] The advertising campaigns highlighted the theme of car radio as companion.[21]

In car radio the "modern-day Robinson Crusoe," as one automobile journalist wrote in the mid-1930s, had found his "Friday."[22] Other automobile journalists as well as radio enthusiasts shared this view. Car radio would help out the commuter for whom covering the same stretch on a daily basis had lost every charm (Hellmut 1933: 13). The alliance of car and radio enabled drivers to withdraw from the world *and* be connected to whatever they liked (von Laffert 1931). And once the driver knew how to find the stations, another journalist noted, tuning into them required "no more attention than lighting a cigarette."[23] In 1948, the anonymous

15. PCA, File 811.215, Product Documentatie Autoradio's, "Storingen, veroorzaakt door ontladingen van statische electriciteit in de wielen," 1937: 7.

16. PCA, File 811.215, 1933 ff., "Philips Auto Radio: Geeft ontspanning en vergroot de aandacht voor den weg," September 1935.

17. PCA, File 811.215, DA Multomap, "Met Philips Autoradio geheel Europa binnen Uw bereik," 1933; PCA, File 811.215, 1933 ff., "Van de Veluwe naar Londen in een handomdraai," June 1, 1939.

18. PCA, File 811.215, Product Documentation Autoradios, "Philips Auto Radio Tips," 1936 (5, May–June): Front page.

19. PCA, Philips advertenties, 1960–1969, "Bent u vaak alleen met die streep op de weg?" 1960, *Algemeen Handelsblad* (May 10). See also PCA, File 812.215, photos of ad materials 1934–1960, "Cut a long way short," and Display November 27, 1954, C46528, "Les kilomètres sont plus courts avec un autoradio Philips."

20. PCA, File 811.215, 1933 ff., "Philips radiotoestellen doen hun intrede in de auto!" 1934, *De Auto* [February 15]: 256. See also Alexander (1938).

21. PCA, File 811.215, Product Documentation Autoradios, "Philips Auto Radio Tips" 1937 (2, September–October): 14.

22. PCA, File 811.215, 1933 ff., "Europa tussen de bumpers" [1936], *Auto-transport*: 125–27, at 125.

23. PCA, File 811.215, 1933 ff., Philips' Autoradio, "Is radio in auto's gevaarlijk?" September 1935.

Figure 3.6
Philips advertisement (1936)
Source: Philips Company Archives Eindhoven, File 811.215, "Product Documentatie Autoradio's," Philips Auto Radio "Tips," no. 5, May–June 1936, cover page.
Courtesy: Philips Company Archives.

author of the British *Autocar Handbook* did not dare give a "definitive statement" on the distractive nature of car radio, "for so much depends on the driver concerned," but "it can at least be said that car radio is certainly no more distracting than a talkative passenger" (Anonymous 1948). In the United States, Crosley even transformed this idea into a major selling point. One of its ads shows a car with a man at the wheel and a woman in the backseat:

> "Look out!" "Be careful!" "Ooh! We're coming to a hill!" No longer need you grit your teeth and suffer such well-meant but irritating exclamations. Instead—just turn on your CROSLEY ROAMIO Automobile Radio Receiving Set. All fear and irritability are forgotten as you spin along the smooth road enjoying concerts, symphonies, the latest dance hits, or the wit of world-famous humorists. (Crosley Roamio radio advertisement 1930, reprinted in Matteson 1987: 80)

Outside the city, as automotive journalists had it, listening to car radio was not so much "dangerous" but actually prevented the driver from entering a "stupor," as was also true of smoking a cigarette, opening a car window, or pausing to take a nap (Strepp 1964: 145–48; Dillenburger 1957: 411, 424–25; see also Peppink and Swanenburg 1954: 117; Peppink 1956: 10; Berk 1961: 209).[24] Alexander Spoerl, the German automotive journalist we cited above, considered "daydreaming" much more dangerous than car radio. While attention to the road might be affected by people's thoughts, that effect was less likely for radio programs, as programmers made sure that it would not happen (Spoerl 1963b: 150).

A study of Swiss car radio owners in the mid-1950s underscored that the radio's companionship on long trips was indeed important to consumers. More than half of the owners indicated that "entertainment" during "long stretches" was the main reason for purchasing a car with a radio. Music was by far the most popular fare among listeners.[25] In 1954, the editor in chief of a leading Dutch motorist journal and his coauthor claimed that a car radio needed to be rather loud to drown out "engine sound" and "street noise" (Peppink and Swanenburg 1954: 118). Against the background of this remark, it is all the more striking that drivers' preference for listening to music was established *before* transistor radio, cassette players, FM reception, and stereo sound became widely available and stabilized the quality of mobile sound (Anonymous 1957; Anonymous 1967; Anonymous 1969: 14–18; Hinlopen 1971: 174–92; Kierdorf 1970; Matteson 1987: 256–72).

It is clear, then, that Philips adjusted both car radio technology and the rhetoric of sales to existing practices within Europe's mobility culture. With its design of a car battery radio with internal reduction of interference, Philips bowed to automobile manufacturers and created a European car radio through its focus on longwave reception. Likewise, with its ad for radio use in stationary cars the company demonstrated its awareness of social concerns regarding the device's distractive character. The same applied to its marketing of radio as an entertaining companion for drivers in the countryside rather than in the city. Radio

24. PCA, File 811.215, 1933 ff., "Autoradio gevaarlijk op den weg?" 1934, *Het Volk* (September 19); "Voor garage en wagen," 1936, *Auto-transport* (September): 261. See also PCA, File 811.215, 1933 ff., C. Tindal, "Auto en Radio," 1934, *Het Motorrijwiel en de Populaire Auto* (February 16): 196–98, at 197–98, and M. W. H. de Gorter, 1937, "Is Auto-radio Gevaarlijk?" *De Installateur* (April 21): 124–[25]; "Radioklanken rond het stuurwiel," 1937, *De Auto* (April 1).

25. PCA, File 811.215, Autoradios, map 1 (1947–1974), *De markt voor autoradio's in personenauto's in Zwitserland*: 2, 16–17. See also "De autoradio en de gebruiker," 1957 (January 2): 6.

lovers, automobile journalists, and consumers shared or took over this perspective. Car radio was thus domesticated into a sound and safe companion on the road.

THE CAR RADIO AS MOOD REGULATOR

The meaning of car radio, however, would hardly remain stable over the years. Manufacturers such as Philips introduced car radio as a way to *draw people into* cars, both literally—attracting admirers—and virtually: the company of radio voices. From the 1960s, however, car radio was increasingly presented as a device assisting motorists to tolerate their fellow traffic participants and keep them, emotionally, *at a distance*.

The traffic system in Western countries changed dramatically between the late 1930s and the 1960s because of the explosion of car use. In 1960, the US Census showed that 64 percent of all employees used either a private automobile or a carpool in order to get to work. A dissertation on commuting that repeated these data opened with the claim that in "most of the large United States metropolitan areas, a monumental traffic jam signals the arrival of the work day" (Aangeenbrug 1965: 7, 1). At that time, car ownership and commuting by car were less widespread in Europe. But even in Europe, urban congestion became a major issue from the late 1950s onward, partly owing to Europe's age-old city structures that accommodated the automobile slowly (Lundin 2008: 263–64).

The late 1930s survey by the American radio researcher Edward Suchman appeared to indicate that radio might alleviate the frustrations this congestion created in motorists. One of the arguments brought forward by drivers and taxi drivers in favor of car radio was that it "[r]educes impatience and annoyance with traffic." When being entertained by the radio, the driver "does not mind waiting" for the green light and drives "[s]teadier, easier and safer" when dealing with "the hazards of cutting in and out of traffic lanes" (Suchman 1939: 154). This theme returned in advertising in the 1960s, as is nicely illustrated by a 1963 animated commercial from Philips aimed at the American consumer. In a "not that way, but this way" scenario we first see a motorist worming his car through dense traffic. He is in an exceedingly bad mood and has an aggressive driving style, raging about others on the road. Things can be much better, the commercial argues: a Philips car radio allows the motorist to move through traffic politely and in good company as well as good spirits.

Figure 3.7
Philips draft advertisement (1956)
Source: Philips Company Archives Eindhoven, File 812.215, Advertisement material 1934–1960.
Courtesy: Philips Company Archives.

For when you listen to your favorite programs, it is very easy to stay calm: "Your fellow-road users will love you, and you will love ... Philips."[26]

Ad materials conveying this message were also used in the Netherlands and other European countries. These ads in the 1960s stressed that driving in a bad mood meant driving unsafely: "You cannot feel at ease everywhere or with everyone," as a brochure in English put it, "especially in today's tiresome traffic—slow, halting, and enervating as it is." A good mood was essential, and car radio brought it within reach.[27] In the mid-1950s Philips already imagined car radio as an umbrella of music, offering protection against figurative rain showers (figure 3.7).[28] As argued in the previous chapter, such intimacy had been evoked as early as the 1920s, when manufacturers started to present their cars as living rooms on wheels (see also Marsch and Collett 1986: 11; Möser 2002: 240; Mom 2008b). But in

26. PCA, "Stop ... go," video, 1952.
27. PCA, File 812.215, photos of ad materials, 1934–1960, "Philips all-transistor car radio," 1963. See also PCA, File 811.215, Product Documentation Philips Autoradio, "Gute Laune unterwegs," 1964.
28. PCA, File 812.215, photos of ad materials, 1934–1960, draft sketch ad, 1956.

the 1960s, car radio was added as a personal mood regulator. Drivers who listened to the radio would be able to cope with the outside world and "reconcile" themselves with it.[29]

An essay published in 1953 by the Dutch psychology professor David Jacob van Lennep tried to clarify the psychology of driving in phenomenologist terms, a dominant tradition in Dutch psychology at that time (Dehue 1995). Van Lennep's context-bound analysis of the world of the driver—his car, the road, his fellow motorists—offers an insightful interpretation of driving in the mid-1950s, and helps *us* to understand how much it changed when roads became increasingly crowded. The peculiarity of the "physiognomy of the road situation," van Lennep explained, was that the driver was at once solitary in the "private envelope" of his car, and at the very same time genuinely social, in continuous "conversation" with his fellow drivers, whom he had to observe, interpret, and understand. Traffic rules structured their communication: good drivers spoke a common language and "listened" to each other carefully. "The one who cannot listen is usually a bad driver," van Lennep claimed (1953: 158–60).

This did not imply, however, that the driver was the same person he was at home. Good drivers were one with their cars, the car being part of the body—and the good driver was supposed to be able to anticipate possible scenarios, taking into account that drivers commit errors. Essentially, however, driving also had an isolating effect on the driver. "Sitting in the small cozy glass room, *separated* from the rest of the world and looking outside through a *window*, I am considered to act courteously towards other cars, whose drivers I cannot see and even ought not to see, but whom I should *imagine*" (van Lennep 1953: 165). The driver needed to let these invisible fellow citizens pass, or should pass *them* as politely as possible, and stay behind them, give them the right of way, or stop for them. This meant that "using roads together" always implied something "negative," which did not remove one's isolation, but actually underlined it. One encountered fellow drivers as correctly as possible and left them behind as soon as possible. It was a conversation without "engagement," an introduction without the promise of getting to know each other. This was what made driving a bit "contradictory" and "ambivalent" if not "unreal." It also clarified why driving bumper cars at the fair was so much more fun—here one deliberately sought contact with fellow drivers (165–66).

29. PCA, File 811.215, 1933 ff., *Philips Autonieuws*, February 1965; PCA, File 811.215, 1933 ff., "Mogelijkheden voor autoradio's" 1966, *Financieel Dagblad* (August 30): 7.

Van Lennep's analysis reminds us not only of Simmel's early twentieth-century take on urban encounters, but also of comments in the 1920s on driving in closed body automobiles discussed in the previous chapter. Simmel underlined the transitory character of meeting each other in urban public space, while the other commentators noted the driver's distanced position in an enclosed car. Van Lennep, however, added the element of the need to constantly "talk" to fellow drivers—in order to avoid accidents—without having intimate contact, at a time when it became increasingly rare indeed that drivers had the road to themselves.

With this in mind, it is possible to develop another take on the marketing strategies behind car radio and its putative functions. Initially, car radio embodied the promise of family entertainment at picnics and of company for the lonely driver—and contemporary studies confirmed that drivers indeed appreciated car radio for its virtual company. Later, car radio was said to reduce the stress of coping with a multitude of fellow drivers, and to enhance one's mood. We do not know of statistics from the 1960s and 1970s confirming such functions, but Michael Bull's work on the 1990s does so for the mood-affecting aspect of car radio (see chapter 1). Moreover, a recent experiment has shown that listening to self-selected music in a car "influences the experienced mood while driving, which in turn can impact driving behavior" (van der Zwaag et al. 2012: 12). Van Lennep's articulation of the peculiar form of communication drivers had to master on crowded roads makes the focus of marketeers on car radio's assistance in coping with traffic even more understandable: car radio, one could say, constituted the fellow driver's voice the driver imagined but could not hear.

Car radio manufacturers had even more in store. Traffic information via radio, as the German industry underlined, offered the driver specific alternatives for action in a world that left few other options for being in control (Weber 2008: 146–49). In the United States, a Newark radio station started broadcasting traffic reports on the basis of data collected by an airplane as early as in 1937, and by the 1950s, "aerial traffic reporting by radio" had become "a regular service during the weekday rush hours in majors cities," now with help of helicopters (McDonald 2008a: 22). French radio stations began broadcasting programs tailored to motorists and their traffic information needs, such as *Route de Nuit*, from the mid-1950s onward (Fesneau 2009: 243). This became a more widespread European phenomenon in the 1960s and 1970s. "Reports on the weather, roads broken up, and traffic jams" were indispensable for the "realistic" motorist, a Dutch handbook author explained. FM stations were particularly active in providing such information (Hinlopen 1971: 174, 181).

Figure 3.8
Blaupunkt advertisement (1939)
Source: Bosch Company Archives Stuttgart, File 1601 001, Pamphlet "Musik im Auto, Musik wie zu Hause," 2.
Courtesy: Robert Bosch GmbH.

In addition, Blaupunkt developed a traffic broadcasting system called Autofahrer Rundfunk Information (ARI) that used a decoder to offer information on the weather, traffic delays, and diversions in whatever German region the driver would pass, a system introduced in 1974 (Weber 2008: 148; Duckeck 1973: 130). Blaupunkt, much like Philips, introduced car radio as one's "most amusing" travel companion in the 1930s, presented it as a heartening "guardian angel" in the 1950s, and transformed it into a pilot guiding the driver through heavy traffic in the 1960s and 1970s (figures 3.8–3.10).[30] We will address traffic radio in our next chapter.

30. Robert Bosch GmbH, Historical Communications (Stuttgart, Germany): Bosch Archives Files 1601 001–1601 092 (1932–1980) on the car radio, and File 1601 623 on traffic information systems. See also Anonymous 2005. For car radio as "most amusing" companion, see Bosch Archives File 1601 001, Pamphlet "Blaupunkt Auto-Empfänger" [1932]: 3; for car radio as "guardian angel," see Bosch Archives File 1601 005, Pamphlet "Schutzengel schweben über uns..." [1950]: 1; for traffic information through car radio, see Bosch Archives File 1601 053 (1967–1968), Pamphlet "Blaupunkt Car Radio" [1967–68]: 1.

Figure 3.9
Blaupunkt advertisement (1950)
Source: Bosch Company Archives Stuttgart, File 1601 005, Pamphlet "Schutzengel schweben über uns," 1950, 1.
Courtesy: Robert Bosch GmbH.

Figure 3.10
Blaupunkt Autoradio, details (1974)
Source: Bosch Company Archives Stuttgart, File 1601 077 I, Pamphlet Blaupunkt Autoradio 1974/1 (1974), 3.
Courtesy: Robert Bosch GmbH.

Here we would like to flag that car radio manufacturers, with traffic radio and other new features, sought to "compensate" for the immobility drivers increasingly had to cope with. It assisted the driver, as Blaupunkt underlined, in spending drivers' time with their families instead of in traffic jams (Weber 2008: 149).

CAR TALK

We have returned here to the point we sketched in the opening scenes of the book: that of the motorist who may acquire, through the sounds of and within his car, a sense of privacy and choice that has come under pressure outside of his car. With the new emphasis on radio as mood regulator, car radio manufacturers, in their sales strategy, responded to the growing traffic density. At the same time motorists, by means of a burgeoning car literature, were taught a new auditory culture. They were supposed to listen to their radio to keep up their spirits; the various sounds and noises of the car's functioning no longer mattered that much. While the automotive industry *dis*couraged drivers from listening to their engine, radio manufacturers *en*couraged motorists to listen to car radio. Drivers thus had to delearn one form of listening and learn another, to delisten and relisten. It was as if the volume control of the engine was turned down, and that of car radio turned up.

This last shift did not stop with the marketing of car radio as mood regulator. On the contrary, listening to the music and sounds of one's choice in the car, creating a fully artificial, sonic refuge from inside the car, would take two additional decisive turns. One was related to the rise of the corridor, not only in the meaning of limited-access road, but also as a tunnel-like environment, the highway that leads the motorist to a destination while giving less and less away of the surrounding landscape. The second was the introduction of audio systems beyond car radio, such as cassette players, compact disks, MP3 players, and the mobile phone.

Before we will discuss these phenomena and their consequences for acoustic cocooning, however, we would like to slightly qualify our remarks about the process of "delistening" to cars. Certainly, German car mechanics secured diagnostic listening as their exclusive skill in the 1930s, and in the United States a long-lingering repair crisis stimulated the introduction of measuring instruments that showed instead of "told" the result of examining the car—a development that also took root beyond the United States, notably after World War II (Krebs 2013, forthcoming). And indeed, even monitory listening by drivers acquired a different meaning over time, as we have seen: it could not only mean "Go and see your mechanic," but also "Why not buy a new car now that this one is bothering you?"

The practice of listening to one's car, however, has never completely subsided, as our audiences informed us when we presented earlier versions of this chapter's argument. As they assured us, they would certainly worry and check after hearing some odd sound from their car. The popularity of the American radio show *Car Talk* is relevant here. This NPR program, which has been aired since the late 1970s, provides additional proof of the ongoing significance of listening to cars. In this one-hour weekly show, callers phone in to receive advice from the two hosts—expert mechanics—about some car trouble they have. Quite often the callers begin by referring to some strange sound they hear while driving that worries them. Usually they do not just describe such a sound ("like an amorous couple under the hood"), but many also take a shot at mimicking it. During the phone call, the two hosts subsequently work toward formulating a diagnosis of the problem, usually by asking the callers more detailed questions about when or where they can hear their car generate the sound that troubles them. In a recent show, a driver complained about his "manly" car's "unmanly" sounds—"it birds, it chirps"—when going over a bump. At the repair shop the car failed to make the odd sound, of course, as the driver tells us with a sense of humor and frustration. In the end the hosts once again manage to solve the mystery: the car has a broken stabilizer link.[31]

In a nutshell, *Car Talk* captures the gist of our argument here: as experts, the hosts do the diagnostic listening; as drivers, the callers do the monitory listening, while the members of the audience enjoy listening to the show on their radio, at home or in their car. Over the years, in other words, motorists have kept listening to their car indeed. As is true of us: we keep listening to cars in the next chapter, and, in fact, to the environment they pass by.

31. See http://www.cartalk.com/Radio/WeeklyShow/online.html, segments 5 and 10, for instance. We accessed these segments in August 2011. They are no longer available online.

CHAPTER 4
"Like a Boxed Calf in a Traffic Drain"
The Car on the Corridor

Before I built a wall I'd ask to know
What I was walling in or walling out,
And to whom I was like to give offence.
Something there is that doesn't love a wall.
From the poem "Mending Wall,"
Robert Frost, *North of Boston*, 1914 (p. 12)

SERVING THE EAR

In January 2008, Marc Eijbersen, project manager at CROW, a Dutch National Knowledge Platform for Infrastructure, Traffic, Transport and Public Space,[1] announced a new directive on noise-reducing constructions along highways. His first sentences suggested great enthusiasm. "Adopt it and you'll be up-to-date again." He promised news about issues such as noise barrier top edges, integrated noise barriers, and modular systems, and ended his first paragraph with a near-cliffhanger: "There is an entire world behind this seemingly simple provision." The tone of his next sentence, however, was quite dissimilar, indicating that nobody, in fact, wanted any barriers along the highway, or, as he put it, nobody enjoyed "driving like a boxed calf in a traffic drain." Who would, indeed, like to be deprived of blue skies, green meadows, and dappled, white, or grayish low clouds—vistas

1. Originally, CROW was an acronym for Centrum voor Regelgeving en Onderzoek in de Grond-, Water- en Wegenbouw en de Verkeerstechniek. It is now considered a stand-alone name.

so common when driving through the Dutch landscape? Yet, Eijbersen stressed, due to laws that protect people who reside near highways against traffic noise, motorists in the Netherlands were likely to have to put up with roadside noise barriers until noiseless vehicles and sound-deadening road surfaces rendered the noise barrier an obsolete technology (Eijbersen 2008: 2).

The announcement by Eijbersen beautifully laid bare the societal and sensorial dilemmas underlying noise barriers. The barriers have been designed to screen citizens from noise, but they also obstruct drivers' views. They serve the ear, but often seem to neglect the eye. Whereas highways promise freedom of movement, the noise barriers along them are reported to have a hemming-in effect on drivers. If the barriers provided nearby residents with protection against noise, the very source of the noise is simply left unaddressed. And while the barriers may seem to be straightforward objects, the science and technology behind them have proved to be rather complex and controversial.

This chapter unravels the history of noise barriers along motorways since the 1970s, and takes the dilemmas just mentioned as its starting point. What were the contexts in which noise barriers were presented as a useful technology? What affected their design and subsequent development? How did drivers assess their presence? How did architects, engineers, and authorities react to public debate about the constructions? Our analysis starts in the Netherlands, one of the most densely populated and motorized countries in the West, and internationally recognized as the country most advanced in roadside noise barrier design (Kotzen and English 2009: ix).[2] We will further contextualize the Dutch situation by taking into account the history of the corridor and its noise barriers in other European countries and the United States.[3]

2. With 7.6 million passenger cars as of January 1, 2010 (about one per two inhabitants) and traffic jams that may cover half of its highways on a winter morning with bad weather, the Netherlands has been an excellent site to explore the history of the highway noise barrier (http://www.cbs.nl/nl-NL/menu/themas/verkeer-vervoer/nieuws/default.htm, accessed September 9, 2011). We have studied documents published by Dutch road authorities, proceedings of Het Nederlands Wegencongres (an association that fostered a national system of highways), journals and other publications issued by organizations of drivers (such as *De Kampioen*, which has a digitized corpus 1885–2010, searched with the help of keywords), civil engineers (such as *Wegen*, digitized corpus, 1980–2010, searched with help of keywords), architects, acousticians, and manufacturers of noise barriers.

3. To contextualize the Dutch situation, we have studied British, German, and American guides for the design of noise barriers as well as publications by the US Transportation Research Board.

We argue that we can best grasp the contested character of the noise barrier by seeing it socially as a specific type of NIMBYism—with the driver considering the landscape the private garden for his mobile home—and culturally as an expression of a deep-rooted fear that bears a remarkable relation with a sense of awe for the "underground" (Williams 2008 [1990]). Drivers clearly favored noise barriers whose designs not only reckoned with environmental concerns, but also considered the trope of the unfolding garden and the motif of an imagined escape from the underground. In general, however, the many miles of noise barriers—notably in densely built-up, urbanized areas—were seen as a sensory deprivation forced upon drivers; a deprivation for which they could only find compensation *within* their cars. Car radio and stereo sets took on new functions and meanings: radio broadcasts provided highway traffic information and drivers increasingly used their audio equipment to play audio books—new forms of acoustic cocooning that helped drivers regain control, or a sense of control, on a corridor that largely controlled them.

THE "ONLY OPTION": THE RISE OF THE ROADSIDE NOISE BARRIER IN THE WESTERN WORLD

Roadside noise barriers bear different names. In the United States the terms "highway noise barriers" and "traffic noise barriers" are used; there's the more recent UK term "environmental noise barriers," and "acoustic noise barriers," the English translation of the term *geluidsschermen* the Dutch came up with. In the early 1980s, the US Transportation Research Board defined the highway noise barrier as a "noise abatement device with a mass and geometric configuration sufficient to provide transmission loss and diffraction of noise propagating from a highway to a receptor" (Cohn 1981: 3). Washington State boasted the first in the United States: it acquired its earliest noise barrier, an earthen wall, in 1963, along State Route 520 near the Evergreen Point Floating Bridge (Cohn 1981: 50; Sullivan 2003: 10).

The theory behind the noise screens had withal a much longer history. It had been known for quite some time from everyday practice that large structures created "acoustic shadows" right behind them. In battle, soldiers hiding behind hills did not hear the enemy approaching and consequently suffered defeat (Ross 2004: 270). In the 1940s and 1950s, physicists used theories of diffraction that were developed for optical waves to account for the behavior of sound waves attenuated by screens. However, the problem with these first models was that they situated both the source and the receiver on the ground, neglecting the effects of ground reflections and the actual height of the source. It was acoustics engineer Zyun-Iti Maekawa at Kobe University, Japan, who calculated and measured how these factors affected the "insertion loss" of barriers: the difference in sound energy at a particular position without and with a noise screen (Maekawa 1968; Kranendonk 1973: 2–4).

In the 1970s and 1980s, acousticians kept fine-tuning the models used to compute the redistribution of sound energy by noise barriers, for instance by distinguishing between "a *diffracted* path, over the top of the barrier; a *transmitted* path, through the barrier; and a *reflected* path, directed away from the receiver" (Simpson 1976: 2-1). They determined the transmission loss values of barrier materials such as wood, steel, and concrete (given the spectra typical of highway noise) and measured to what degree attenuation depended on the frequency of sound waves. Barriers built in parallel at both sides of the road happened to create bothersome multiple reflections, so the acousticians calculated the improvement that barriers with *sloped* sides could generate. A barrier attenuation of 10 decibels meant a reduction of the sound level by half—attainable for noise barriers that were not too high, say a height of less than five meters (Simpson 1976: 3–6; Kranendonk 1973: 5–7; Rijkswaterstaat 1979: 9). Acousticians also began criticizing some of the assumptions in Maekawa's model. While Maekawa had made his calculations while using screens of semi-infinite lateral length, the measured insertion loss with finite screens was much lower, at least for some source-receiver positions. Moreover, Maekawa's figures were said to work only for screens that were very thin and had razor-shape profiles. Thicker profiles diffracted the sound waves "at both edges of the rectangular profile," and rounder ones actually carried rather than blocked the sound waves. In the 1980s and 1990s, studies of profiles thrived, among which were the double thin wall, the L-shaped barrier, and the controversial T-shaped screen (Davies 1994: 22–24; Watts, Crombie, and Hothersall 1994; Samuels and Ancich 2002).

Yet it was not so much the expanding knowledge behind the noise barrier that influenced its rapid rise along highways, as the introduction of a new approach to antinoise policies in the 1970s. In the West, noise had been on the agenda of national governments since the late nineteenth century. Over time, governments had introduced noise abatement regulations that controlled the sound level of particular sources such as factory machines, passenger cars, and aircraft. At the end of the 1960s, however, they started to see noise as an environmental concern that demanded a more encompassing approach. Rather than tackling the noise *emission* of each source individually, the added noise "*immission*" that citizens had to cope with became the focus of attention. As a consequence, some years later, framework laws were passed that brought different kinds of noise and noise control under one heading (Bijsterveld 2008). The 1960s and 1970s were also the decades in which acousticians and committees advising governments on noise produced a wealth of publications with statistics about the predominance of road traffic noise in the overall noise pattern in urban areas. The widely cited Wilson Report, commissioned by the British government and published in 1963, was one of these (Wilson 1963: 22, 27). French, German, Swedish, Norwegian, and US surveys produced similar findings (Alexandre 1974; Kihlman 1975; Kirschhofer 1970).

It was in this context that noise barriers started to attract growing interest from policy advisors and makers in Western countries. The German government, for instance, declared noise to be the focus of its environmental policy in 1976, regulated protection from road noise in its *Bundes-Immissionsschutzgesetz* of 1977, and secured annual funding for "noise renovation" of roads in 1978, recognizing the relative importance of *Sofortmaßnahme*, or short-term solutions such as noise barriers (Krell 1980: 17–18). Federal legislation in the United States similarly started to allow "funds to be used for voluntarily 'retrofitting' existing highways with noise barriers" (Cohn 1981: 5). It should be noted that other means of noise abatement like maximum noise emission levels for cars, urban planning, traffic regulation, and sound-deadening asphalt did not lose their importance—we will take up that story in the next chapter. Yet for heavy traffic in urban areas, noise barriers seemed to promise a quick solution. In 1973, the Dutch Interdepartmental Committee on Noise Nuisance (Interdepartementale Commissie Geluidhinder) published its first report on noise barriers. Its opening statement was revealing: "In densely populated areas, it is not always possible to create the distance between dwellings and main roads that is necessary to prevent noise nuisance. The only option to prevent nuisance, then, is to mask sound with the help of planting, entrenchments, earth walls, or soundproof screens" (Kranendonk 1973: 1).

The belief that the noise barrier was the ultimate solution to transportation noise in urban settings and the bringing into effect of the Dutch Noise Nuisance Law in 1979—setting a maximum noise immission level of 55 dB(A) for residential areas near existing roads—spurred the rise of roadside noise barriers in the Netherlands (Rijkswaterstaat 1986: xiii). Between 1975 and 2000, the Netherlands acquired a series of noise barriers along its state highways with the accumulated length of 280 miles. The second half of the 1980s were the years that saw the highest growth rate, and between

2000 and 2007, the Dutch added another annual average of 12 miles.[4] The Netherlands created a much greater density of noise barriers than, for instance, the United States. By 2007, the United States had 2,506 miles of highway noise barriers,[5] compared to 367 miles in the Netherlands. Bearing in mind that the United States had somewhat more than 46,700 miles of limited access roads[6] around that time, whereas the Netherlands had about 1,490 miles,[7] the Dutch had a significantly greater density of noise barriers (more than 24 percent versus over 5 percent). Moreover, the Dutch could not adopt the American option of embedding noise barriers in the rolling countryside (Simpson 1976: 3–37), because the Dutch had no hills available. This made noise screens highly conspicuous elements in the landscape. Despite such differences, however, the most *urbanized* areas in both countries acquired the highest density of noise screens. In the United States, for instance, the state of California had the highest number of barrier projects, and Los Angeles was the metropolitan area with the most noise barriers (Cohn 1981: 5). This in turn affected the experience of driving in these areas.

Even though the Americans were the trailblazers in noise barrier construction, authors of US publications on barrier design increasingly turned to Europe for inspiration, and expressed their admiration for what they considered the daring and standard-setting designs of the Europeans (Billera, Parsons, and Hetrick 1997: 55). The second edition of the UK *Environmental Noise Barriers: A Guide to Their Acoustic and Visual Design* indicated the Dutch as the leaders in noise barrier design. "As before," the authors Benz Kotzen and Colin English claimed, "it is the Netherlands that leads the way in barrier development and innovation" (2009: ix). The Dutch

4. CBS, PBL, Wageningen UR (2007), Geluidsschermen en ZOAB in Nederland, 1975–2000 (indicator 0405, draft 03, August 31, 2007), at http://www.compendiumvoordeleefomgeving.nl/indicatoren/nl0405-Geluidsschermen-en-ZOAB.html?i=16-44 (accessed September 9, 2011). See also Rijkswaterstaat 1989a.

5. "Summary of Noise Barriers Constructed by December 31, 2007," at http://www.fhwa.dot.gov/environment/noise/noise_barriers/inventory/summary/stable75.cfm (accessed September 9, 2011).

6. By 2009, the United States had a total length of about 75,000 km (over 46,700 miles) of limited access roads; see http://en.wikipedia.org/wiki/Transportation_in_the_United_States (accessed September 9, 2011).

7. http://www.wegenwiki.nl/Nederland (accessed September 9, 2011). It is difficult to compare the situation in the United States and the Netherlands, as the figures do not refer to exactly the same things. Yet the Dutch highways (*snelwegen*), usually maintained by national road authorities, come closest to the US limited access roads (state or interstate). For both countries we have taken the total length of noise barriers, even though some of these barriers may have been built along roads other than limited access roads. We would like to thank Thomas Zeller for explaining US road terminology.

had been able, for instance, to create barriers that not only blocked noise but also reduced air pollution, and to integrate noise barriers with "non noise-sensitive buildings" (2009: x). An earlier edition of this guide, published in 1999, also took most of its examples from the Netherlands, and to a lesser extent from other European countries (Kotzen and English 1999). It mentioned only a few examples from Japan, even though the Japanese had contributed significantly to noise barrier theory, were highly active noise barrier builders, and started to develop active noise-control devices (such as neutralizing sound waves with counterwaves) for barriers at the end of the twentieth century (OECD 1995; Samuels and Ancich 2002).[8] At that time, the Dutch still eschewed this technology (Ministerie 2008: 16). It is safe to say, however, that among the European and North American countries, the Dutch stood out in their enthusiasm for noise barriers and experimenting with their design—investing in them as if they were, as several commentators would have it, their new dikes.

A NEW BERLIN WALL: THE RECEPTION OF NOISE BARRIERS IN THE NETHERLANDS

By the early 1980s, Dutch road authorities and experts began to acknowledge that the noise barrier was not exactly *the* staircase to the heaven of silence. The barriers came with a price, not only literally, but also in terms of disadvantages such as visual hindrance. The US Transportation Research Board claimed that the effectiveness of noise barriers as perceived by residents and motorists was "often more influenced by the aesthetics and landscaping of a barrier than by the acoustical performance" (Cohn 1981: 1). Over time, even the wording of the visual and aesthetic problems related to noise barriers acquired an increasingly negative tone. While many US publications talked about potential "visual hindrance" in the 1980s, this had shifted to "visual pollution" by the mid-1990s (Storey and Godfrey 1996: 107).

Most experts worried more about aesthetic objections against barriers voiced by residents—to be accommodated through public hearings—than about comments by drivers.[9] This stance may have been prompted by the simple fact that motorists actually did not often share their assessments of

8. See also Reducing Traffic Noise with a Noise Barrier that Uses Sound http://www.japanfs.org/en/pages/025597.html (accessed September 12, 2011).

9. Herman Otto has, for instance, claimed that the road side of the noise barrier initially received less attention than the residential side (Otto 1985: 401). For reports fully focusing on the residents' perspective, see Kortbeek and de Boer (1987) and Ministerie (1989).

the barriers. A 1980 survey on the experiences of US state highway agencies with noise barriers mentioned that "[o]nly two states noted negative driver reaction to in-place barriers" (Cohn 1981: 11).[10] And a Dutch report on noise barriers claimed that drivers did not like the barriers but rarely expressed their negative evaluation (Vereniging 1991: 27).

In a 1986 Dutch questionnaire on their experience of highways, however, both residents and drivers spontaneously expressed their views on roadside noise barriers. As the researchers stressed repeatedly, the respondents—partially recruited in service stations along the highways—had not even been *asked* to reflect on noise barriers. Given an opportunity, they *did* comment on the barriers, and what they had to say was hardly positive. The respondents straightforwardly classified noise barriers as objects that made highways unattractive. The report did not always strictly distinguish between drivers' and residents' answers, yet many of the respondents' remarks on highways were clearly expressed from the driver's perspectives—many residents being, obviously, also drivers. "Something there is that doesn't love a wall," the American poet Robert Frost wrote in 1914 (12). Yet what exactly was the "something" for *this* particular type of wall? What made drivers utterly dislike the noise barriers, though they did not always publicly voice their loathing?

First of all, they referred to noise barriers as "fences," or *schuttingen* in Dutch, a word imbued with negative connotations. The term *schuttingwoord*, for instance, is "a term of abuse," an obscenity. These "fences" were unattractive due to their "claustrophobic effect," "oppressing" character, "tunnel" associations, "lack of greenery," or straightforward "ugliness." "While driving [on a road fitted with noise barriers], you are cut off from everything," one respondent claimed. Another regretted that the noise barriers took "away one's view of nature." All in all the respondents could see that the barriers served a purpose, but their design was often too monotonous; they could even be said to be "a dissonant element in . . . the landscape." They interfered with one's idea of "naturalness." Motorists likened driving along the barriers to moving through a "gutter" or "drain," a notion we have already encountered earlier. It caused a "sinking feeling," as one driver explained in more detail, a feeling that made the driver slow down. Another compared the barriers to "blinkers" (Boekhorst, Coeterier, and Hoeffnagel 1986: 107, 27, 98). Even experts employed by

10. Yet this may have been an effect of the road agencies' tendency to define public involvement merely as the involvement of residents; a group also more easy to localize for questionnaires than the floating mass of drivers.

Rijkswaterstaat, the Dutch Waterway and Road Authority, increasingly expressed the driver's perspectives in these terms. While the "obstructed" visual contact between the drivers and their surroundings was considered a potentially attention-undermining and oppressing "hindrance" in 1979 (Rijkswaterstaat 1979: 12), almost twenty years later, regular road users were said to be "imprisoned" in a "tunnel system" (*gangenstelsel*) of soundproof constructions (Padmos, de Roo, and Niewenhuys 1998: 28). One 1986 respondent disliked the barriers enough to want "to put the noise screen hype on hold," had the respondent had the power to do so (Boekhorst, Coeterier, and Hoeffnagel 1986: 107).

Four interconnected yet distinct themes were predominant in drivers' negative evaluations of noise barriers: the claustrophobic effect, the beneath-the-earth experience, the gutter feeling, and the involuntary seclusion from the natural landscape. The noise barriers had an imprisoning, enveloping effect that many drivers also claimed to experience in tunnels. They clearly associated the imprisoning effect with being underground, as if a fenced road lowered itself, even though this was often only an illusion. The references to "gutters" or "drains" expressed not only the experience of driving through a tunnel, but also one's lack of control—as if one was immersed in a river that dragged the driver along. These comparisons expressed the same lack of agency as did the metaphor of the "boxed calf" and of a horse with "blinkers." Ultimately, the noise barrier created an artificial environment that contrasted with nature and the rural landscape.

These complaints partially overlapped with drivers' complaints recorded in the United States in the early 1980s. In Virginia, drivers said that noise barriers made "driving monotonous," blocked "the scenery," and were simply "eyesores." In Minnesota, the laments came not from commuters "but, surprisingly, from tourists." Unlike Dutch drivers, however, the Americans also expressed concerns about the acoustic effects of the noise barriers and their costs. Judging by their ears, the noise barriers made "the highway ride noisier," probably an effect of the reverberant buildup caused by parallel noise barriers. They were further concerned about the cost effectiveness of the barriers and objected: "Why spend so much on so few?" and "Why do *I* pay for someone else's comfort?" (Cohn 1981: 11). The Dutch, on their part, were not preoccupied with cost-benefit analyses. To them, at that time, paying taxes and public spending were still facts of life that they had to take in their stride. Yet, being fenced off from the scenery—the picturesque nature view—was a concern the Dutch and American drivers shared, at least the ones who had objections to noise barriers.

We will contextualize these concerns about the driver's visual seclusion from nature below, but we will first analyze the trope of being locked-in.

Several articles in Dutch drivers' magazines extensively elaborated on the negative effects of noise barriers and reported dramatic stories. One driver, having engine problems, felt forced to park his car on the highway's hard shoulder. He decided to leave the car and enter a safety door in the noise barrier alongside the highway to look for help. He soon discovered, however, that behind the door there was nothing to assist him. He found himself in a no-man's-land. The worst was yet to come: he could not return to his car since the door had locked itself behind his back. He was "Trapped like a rat," as the article's title had it (Anneveld 1996). In a similar narrative, a woman discovered she had been trapped while her partner and two young children had remained in the car. As she was waiting behind the barrier, her imagination ran wild, producing visions of her partner leaving the car and entering the space behind the barrier to look for her. "I could vividly picture it: my four-year-old son unbuckling himself and stepping out of the car, right onto the A9" (Meijer 2000: 7). She had no mobile phone—a tool that had not yet become ubiquitous in the 1980s and early 1990s. From the roadside, the noise barriers' safety doors were easily accessible and closed automatically. Yet at the other side, the doors could only be unlocked by means of keys kept by road service staff (Goudriaan and van Dool 1983: 57).

Even though such incidents were reported on in the consumer television program *Kassa*, it was not until the beginning of the twenty-first century that the Dutch Road Authority added signs to the doors in noise barriers that spelled out they could only be opened from the road side (Anneveld 1996: 31; CROW 2007: 42). Until that moment feeling "locked in" was not only a metaphor for the negative experience of driving highways fenced off by noise barriers, but being "locked out" by noise barriers or "locked in" in case there were no safety doors at all was also a physical possibility. In practice it was, fortunately, quite uncommon. The author of a German report posited that he did not know of any case in which escaping or acquiring help through service doors had been the only option to find safety. Emergency services taking the highway usually arrived at the scene fast enough (Krell 1980: 248). The stories thus articulated the *feeling* of being locked in and locked out rather than expressing the experience of actually *being* trapped on either side of the barrier.

The significance of this feeling was also raised in a report on noise screens commissioned by the Dutch road authorities and published by a Rotterdam-based urban planning consultancy firm in the mid-1980s. The report summarized the results of a study on the perception and evaluation of video projections of six types of noise barrier by forty-four road users. The results took the researchers by surprise. According to design theory, a noise barrier had to have appealing yet nonstriking colors, should not be conspicuous, had to be partially transparent, needed some variation in appearance, and required planting (Lever 1985: 33; Aarsen 1985: 416). The outcomes confirmed the widespread appreciation of walls with greenery, the aversion to plain concrete barriers, as well as the need for diversified, nonboring barrier design. Yet transparency only worked if applied to long stretches of noise barrier. Apparently, the speed of the driver had insufficiently been taken into account (Lever 1985: 26–30). Harder to interpret, however, was the appreciation for a barrier that was very conspicuous and highly artificial—sporting no planting whatsoever. It had transparent parts in the form of waves that alternated with stretches that displayed colorful horizontal strokes in bright colors. Bright colors were thus more important than theory had assumed—barriers shouldn't be gray and grim, and a striking appearance happened to be quite acceptable. Yet the study also mentioned that this particular barrier stood at a considerable distance from the road, had a slight angle of disinclination, and colors that were lighter at the edge of the barriers than near the road—all of which had been introduced to avoid the "tunnel effect" (Lever 1985: 21). Apparently, this effect was even more relevant for the definition of design criteria than had been imagined.

Another study, again commissioned by the Dutch road authorities, confirmed this finding with the help of perception psychology. For the driver, slanting screens not only had fewer acoustical disadvantages than upright ones, but also fewer visual disadvantages. First of all, they resulted in less blocking of landmarks than the traditional upright noise barriers did, and thus caused less visual obstruction. In addition, the slanted screens created less *asymmetry* of the speed vectors perceived by drivers, and fewer *conflicting* speed vectors. In case of upright barriers, the information from speed vectors below and above eye level conflicted—the ones above eye level pointing upward, yet the ones below eye level pointing to the ground, thus giving drivers the feeling that they were approaching a screen, which instigated a sense of danger (figure 4.1). This phenomenon was less manifest in case of slanting screens. Differences in the perception of road curves, as well as the distance of the barrier edges, their gradualness, and a few aesthetic issues completed the picture (Leeuwenberg and Boselie 1986: 6–10, 20; Kruysse 1990).[11]

Yet perception psychology could not explain the exact wording and stories used by drivers to express their fears and associations. In fact, drivers' choice of words and story lines display a remarkable similarity with

11. For the effect of noise barriers on how drivers position their cars on the road laterally, see Aarsen 1985: 413.

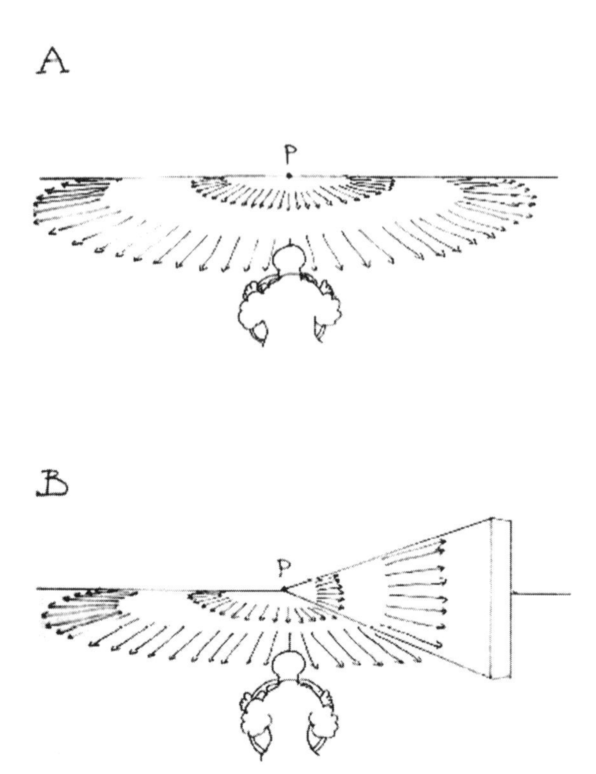

Figure 4.1
Visualization of speed vectors experienced by drivers when passing noise barriers (1986)
Source: Leeuwenberg and Boselie 1986: 20.
Courtesy: E. Leeuwenberg, Radboud Universiteit Nijmegen, the Netherlands.

narratives about the underground. In *Notes on the Underground*, Rosalind Williams explains how between the eighteenth and the late nineteenth centuries the representation and aesthetic evaluation of the underground—the cave, the mine, the tunnel—transformed from an ugly, slimy, noisy and hellish space, via a "dark, deep and deprived" environment evoking a sublime experience of awe and fear, to that of an extraworldly paradise. These shifting aesthetics, which Williams derives from Western literary and nonliterary sources, were linked to technological changes such as the rise of electric lighting in the nineteenth century. Key themes in the late eighteenth and early nineteenth centuries' sublime experience of the underground, for instance, were the "artificial infinite," in which the eye is confronted with nonorganic objects and spaces "of great dimensions," and the feelings of enclosure, unimportance, and submission that came with it. At the end of the nineteenth century, however, sublime awe for the underground gave way to fantasies of fully enclosed, silent, and magic spaces—spaces, as Williams explains, that were as often situated above as under the ground (2008 [1990]: 86, 90).

Interestingly, motorists' descriptions of how they experienced driving along noise barriers echoed both the ancient underground tropes—the bleak ugliness of the oppressing noise barriers and their reverberant noise—and that of the sublime: driving a "barried" road is driving a "buried" road, deprived of visual input, with lengthy barriers of plain concrete that take on the role of the artificial infinite, and the deep drains articulating a forced submission to the high structures along the road. Yet none of the drivers' representations of the road arrives at the phase of the underground paradise. These are, interestingly, only to be found in their appraisal of automotive space, of the car's interior—as we will see below. It is the car itself that brings what Williams calls the "enclosed realm of consumer bliss," a refuge with artificial temperature, light, and sound, a magic carpet compensating for the sensory loss that comes with the noise barrier road (Williams 2008 [1990]: 109). And this is hardly a coincidence. To cite Williams once more: "The more human-made structures degrade the natural environment, the more alluring becomes the self-enclosed, self-constructed paradise. Technological blight promotes technological fantasy." In today's world, Williams notes, we can recognize these magic spaces in the "first-class airplane cabin, the hotel suite, the limousine, the executive office, the fine restaurant, the shopping mall. In all these environments, the world has not so much been disenchanted as reenchanted" (114, 113).

Drivers' laments about the noise barriers were more than imaginations of the underground, however. Their irritation at the loss of natural scenery also reflected the long-standing ideals of the cinematic drive and the living room on wheels—tropes we discussed in chapter 2. At the end of the 1980s, the author of an article in *Kampioen*, a magazine for motorists, claimed that one could still enjoy "[w]oods, heath, meadows, cows, urban beauty even" while driving, but "these fine things are increasingly hidden from the driver's view." The Netherlands was abundant with noise screens, and only the introduction of sound-deadening asphalt might save the countryside (Van der Snoek 1988: 52). Less than twenty years later, however, the editor of the same magazine had lost all hope. Driving simply wasn't fun anymore, and not only because of "those idiots" on the road or because of traffic jams. On the contrary, it was exactly the *quiet* road *without* other drivers that, tragically, could no longer incite pleasure. The scenes of farmhouses and cows, the ditches and canals that reflected the skies, the vistas along groves: all were "spoiled by disgustingly ugly concrete, glass or artful noise screens." Of course, had he lived near the highway, he too would ask for an "isolation fence" to screen off his home. Yet while on the road, the barriers made him drowsy, and he almost preferred a traffic jam to a noise barrier—"who could ever have predicted this?" (Karsemeijer 1997: 5).

As the appreciation of the new situation on the quiet road clarified, the noise barrier threatened the cinematic drive at its very heart, that is, the tourist experience—suffice it to recall the American remark about the Minnesota drivers. Road historians have shown that from the 1930s onward, an important impetus behind road design was the idea that roads and their landscaping should spur tourists' explorations and enhance the appreciation of their own or foreign countries. Even though there were competing views stressing the need for efficient and straightforward transport, the driver's view of the extending countryside was considered of utmost importance by road architects in the United States and Germany, as well as by European tourist organizations (Mauch and Zeller 2008; Schipper 2008; Zeller 2010 [2006]). Yet in 2003, the well-known Dutch architect Francine Houben noted that the view from the car had become a "forbidden view"—there was nothing to see but an "elongated ribbon of characterless prefab offices and noise barriers," creating widespread "discontent" (Kooman 2003: 50). This discontent not only marked the much-regretted end of the cinematic drive, it further engendered an inverted not-in-my-back-yard attitude, or NIMBYism. Drivers understood that residents wanted a noise barrier in their backyards, but they were unwilling to have it ruin *their* backyard, the view they had of the endless gardens along the road from their private castle on wheels. In that sense, their attitude was hardly different from the NIMBYism of people who are confronted with the commanding view from their homes of airports, railroads, windmills, nuclear energy plants and cell phone masts.

Finally, it is interesting to observe that the expressions that indicated a loss of control while driving in a drain, gutter, or "groove" stood in stark contrast to the ideal of "cool" driving that had emerged in the 1920s. As Rudy Koshar has illustrated for Weimar Germany, the advocates of New Objectivity (*Neue Sachlichkeit*) fostered a new, matter-of-fact relationship with the new artifacts of the Roaring Twenties. They particularly celebrated urban traffic as inciting a habitual response or reflex from its participants, such as drivers reacting in a taken-for-granted manner to traffic lights or fellow drivers. Such "cool conduct" helped people go with the flow of "supra-individual" systems and to create "a context for freedom and for concealing the inner tensions that were part and parcel of the modernist sensibility" (Koshar 2005: 125). Drivers' response to noise barriers, however, seemed to signal the limits of "cool conduct," the confinement of merely accepting one's subordinated position in modern systems. Once reflex and flow were replaced by a sense of blindness and helplessness, driving became less cool.

In 2004, a Dutch magazine published a letter to the editor by a motorist who claimed that the Dutch had enthusiastically celebrated the fall of the Berlin Wall, yet were now actually building a wall of their own. An image accompanying the letter showed the Netherlands surrounded by one big noise barrier instead of its famous dikes. The illustration created some interpretative ambiguity, since the Dutch take pride in their dikes. Indeed, several publications promoted this link between dike and barrier-building as signs of advanced civil engineering. The Dutch author of the published letter was far from ambivalent in his appreciation of noise barriers, however. He ended his letter by stating that the time had come to look for alternatives (de Roo 2004: 6). Such alternatives, other drivers and infrastructure experts underlined, had to reduce noise at its source and stop depriving drivers of the Dutch panoramas of outstanding beauty.

DENIAL DESIGN: BUILDING GREEN, TRANSPARENT, AND LOSS-COMPENSATING BARRIERS

It may seem awkward to discuss the reception of noise barriers prior to describing their design—reception is impossible without the artifacts being there—yet a true design discourse of noise barriers only started years after they had been established. Initially, the barriers were treated as a mere acoustic solution to a noise problem, screens as in Zyun-Iti Maekawa's theories. And in contrast to other architectural structures, contractors had acquired a more independent stake in the production processes of noise barriers, since the barriers were seen as rather straightforward products.

Looking back on the early history of noise barriers in the Netherlands, Dutch designer Herman Otto characterized the mid-1970s and 1980s as the years without much "direction" in design. Only in the 1990s did the noise barrier develop into an object of architecture in its own right. Admittedly, before that decade, designers had ideas as to what the noise barriers should look like, but they stressed the need to either visually *deny* the existence of the noise barriers, or to *compensate* for what had been visually lost as a consequence of the barrier (Otto 1991). The two approaches just mentioned are versions of what we would like to describe as *denial design*. The first version aimed to make the barrier itself invisible to drivers and residents by using, for instance, plantings that would overgrow the barrier, or by using natural material such as willows to construct it (Alberts 1985). Camouflage was the key word in this approach (figure 4.2). The second type of approach

Figure 4.2
Green noise barrier (MW-Groeischerm Stijlenmodel), Provincial Road Geldermalsen, the Netherlands (Mostert De Winter)
Source: http://www.mostertdewinter.nl/sites/www.mostertdewinter.nl/files/geldermalsen-geluidscherm-1.jpg, retrieved September 3, 2012.
Courtesy: Mostert De Winter BV, info@mostertdewinter.nl.

sought to compensate for the invisible landscape, either by making the barrier partially transparent, through use of glass or synthetic material, or by including on the barrier a representation of the lost view. Some barriers had concrete surfaces that mimicked "natural" wood, color schemes that evoked green meadows and blue skies (green close to the ground, blue at the top edge of the screen), or wavy forms conjuring the lake behind the screen. One barrier had "noise trees": abstractly designed trees in front of the actual barrier that compensated for the woodlands behind it (Otto 1985).[12]

Denial design was not restricted to the Netherlands. As early as 1976, the US Federal Highway Administration's Department of Transportation published a report that urged extensive landscaping. "Walls should, as much as possible, and where desirable, reflect the character of their surroundings. Where strong significant architectural elements occur in close proximity to wall locations, a relationship of material, texture, and color should be

12. In his 1985 article, Otto had the architectural phase start somewhat earlier, in the mid-1980s, as several other overviews did; see Aarsen 1985: 415; Kranenburg 1985: 420–21; Kranenburg, Otto, and de Vrede 1987.

explored" (Simpson 1976: sec. 3, 29–33). And the 1981 Transportation Research Board report mentioned that in New Brighton, Minnesota, "a sailing and swimming scene" had been depicted on the driver's side of a barrier. "The barrier," the report said, "is protecting a recreational lake, and the state highway agency deemed it essential to maintain the general flavor of the area through the use of murals" (Cohn 1981: 11). Time and again, reports stressed that the appearance of barriers "should reflect the character of their surroundings as much as possible" (OECD 1995: 110).

Even though the 1976 American report referred to the option of counterbalancing the loss of the cityscape, both American and Dutch publications were especially concerned with restoring the broken link with nature. The materials used to construct the noise barriers had to "harmonize" with the barriers' "environment," the Dutch road authorities stated at the end of the 1970s, which implied both urban and rural surroundings. Yet the illustrations in the report gave examples only of the greenery that "had to make up for" the losses caused by the barriers (Rijkwaterstaat 1979: 12, 21). A few years later, the Interdepartmental Committee on Noise Nuisance repeated the remark about the need for harmony, and stressed that earthen walls were less intrusive on "the landscape" than screens (Goudriaan and van Dool 1983: 73, 8). Noise screens, alternatively, required less space, while plantings could solve the visual problems. The committee added that the color scheme of green and brown hues had to "camouflage" the barriers, that one had to "soften" barriers by integrating them into the environment, and that the perceptual height of barriers could be diminished by using transparent material at the top edge of the screens, building screens on earthen walls, creating horizontal bars on the screens, and planting trees to raise the height (Goudriaan and van Dool 1983: 76, 77). The committee also researched transparent screens made of glass or synthetic material (figure 4.3). It mentioned the dangers of optic reflection, accidents, and vandalism, yet listed solutions such as the use of lamellae to break up reflections (Goudriaan and van Dool 1983: 31–39; see also Krell 1980; and Uittenbogerd 1997).

From the mid-1980s onward, Dutch road authorities strongly advised adding a visual spatial analysis (*Visueel Ruimtelijke Analyse*) to the preparatory phases when building noise barriers. Initially, the ideas behind denial design were practically informed: the design should not be "monotonous," yet still "harmonic" by embedding it in its environment and by creating "visual continuity" (Rijkswaterstaat 1986: 3.4–3.8). It was not until the end of the 1980s that *individualizing* the design was mentioned as an option in addition to denying the barrier, adapting it to or embedding it into the environment (Rijkswaterstaat 1989b: 61). Up to that moment, however,

Figure 4.3
Transparent screen, A5-Westrandweg, Amsterdam
Source: Holland Scherm, Rotterdam, The Netherlands (Jacqueline Paris).
Courtesy: Holland Scherm, Rotterdam, The Netherlands.

the authorities largely responded to and co-created the tropes of the cinematic drive and highway garden.

Nevertheless, it is important to note that the idea of designing motorways that reflected the surrounding landscape was not merely conceived to address drivers' objections to noise barriers. As Peter Merriman explained in the case of England, landscape architects developed a road design aesthetic that was grounded in drivers' experiences of high-speed traveling, embodied movement, and mobile gaze as early as the 1940s and 1950s. Architects felt inspired by the German autobahns and American freeways that, at least in their opinion, were embedded in and had enhanced the existing landscape. British landscape architects stressed that roadside landscaping "must reflect functional, modern principles—of simplicity, unobtrusiveness and a sense of visual flow—and the landscape architect must adopt techniques for maintaining the orderly movements of drivers: planting to improve safety, guide the attention of motorists, screen unsightly views, prevent boredom, reduce dazzle and enliven the scene" (Merriman 2006: 95). Instead of beautifying roads with a wide variety of greenery and colorful flowers—a "horticultural approach" that re-created a typical English country garden alongside the road as if it could be appreciated at the pace of a pedestrian—these architects aimed to appropriate a genuinely mobile perspective in road design. Roads should follow the contours and features of the landscape and create for drivers a middle ground between experiencing distraction and boredom. As the British landscape architect Brenda Colvin had underlined in her 1948 book *Land and Landscape*, the ideal road should break "the mechanical monotony of engine sound and road surface" and keep drivers "alert and vigilant" (Colvin quoted by Merriman 2006: 83, and Merriman 2007: 212). In the Netherlands, a country abundant with noise barriers, denial design had to camouflage the fact that the driver's eyes caught less and less of the landscape, which meant that the landscape had to be *in* or projected *onto* the barrier.

MOVING DESIGN: THE GLOCALIZATION OF THE NOISE BARRIER

From the 1990s onward, denial design largely gave way to approaches that welcomed noise barriers as objects of true architecture in three ways. The first trend was to take the speed and mobile perspective of the driver more seriously than had been done before, and to create barriers that seemed to move *with* the driver. The second was to create consistency and standardization in design along particular corridors or trajectories, and to treat the noise barriers along one highway almost musically as variations on a theme. The third was to "glocalize" the noise barrier in such as way that, despite standardization, drivers were able to acquire their bearings in space and identify the *locations* rather than the *landscapes* they passed by without seeing them, for instance through artworks. We would like to bring these design trends under the designation of *moving design*, since the trends all responded in some way or another to the hypothesized experiences of the driver on the move and intended to "move" the driver in a positive way. One commentator even used the term "speed architecture" (Kranenburg 1985: 421).

It is important to note that denial design was not done away with entirely, yet particular aspects of it, such as the ideal of transparency, blended with a new focus on architectural aesthetics. The *sound tube* in Melbourne, Australia, a largely transparent tunnel "designed to reduce roadway noise without detracting from the area's aesthetics" became a notable example.[13] A similar project was a half-open, rounded, and transparent noise-reducing structure that ran for a distance of one kilometer

13. See http://en.wikipedia.org/wiki/Noise_barrier for this quote and a picture of the Melbourne sound tube (accessed at July 7, 2011).

near the A16 in Dordrecht, the Netherlands—called De Huif, or The Hood (Vreuls 1998). At the same time, the more traditional barriers came under severe attack. In 1999, an anonymous article in *Noise & Vibration Worldwide* drew a quite unflattering history of the noise barrier in the UK. The barriers had originally been built "as unimaginative close-boarded timber fences." But the future seemed bright, now that the UK had recently started to follow "the continental trend to liven up the appearance of its barriers. This also gives more interest and stimulus to motorists, whose lateral vision is constrained by the barriers" (Anonymous 1999: 8).

There was more to it than just livening up the barrier, though. Architects started to rethink what it was like to move along a barrier, and adapted their designs accordingly. "Drivers pass [the barrier] in forty seconds," the architect of The Hood claimed, and "they see it at a speed of 60 miles per hour." Had it been a concrete construction, drivers would intuitively have slammed on their brakes when seeing it—hence the architect's decision to create a fully transparent construction that would prevent a panic reaction (van den Boomen 1997: 43). A famous design was that of The Wall, a building cum barrier construction. Passing the 800-meter barrier would take about twenty-four seconds. It had to become a "visual party" and a "definitive turn" in the presentation of Dutch cities to the driver. The architects made computer simulations of driving along the barrier, and gradually adopted the design in such a way that it enhanced or reduced the experience of speed. The result was a red-colored, wavy design with a "head" that seemed to float and a long, gradually tapering tail (figure 4.4) (Verheijen 2004, 8). It is now commended as a "landmark" and a remarkable experience for motorists: the building-barrier seems to "move like a red-colored flowing and billowing mass."[14] At least one wall meant to reduce noise, the architects apparently assumed, had a chance to be held dear.

Creating an experience of smooth visual transformation also meant that a jumble of swiftly alternating designs had to be avoided. Initially, road agencies had urged architects to strike the golden mean between diversity and disorder in the design of noise barriers, and to alternate designs as a way of preventing monotony (Rijkswaterstaat 1979: 12, 18, 24–27; Goudriaan and van Dool 1983: 73). In the 1990s, however, consistency and rhythm in design became the new buzzwords. As a 1995 OECD report claimed, "it must be avoided that each noise barrier section be considered a particular operation to be studied separately, or even be sub-divided into sub-sections." For the driver, it was and should be "a linear structure," in which horizontality had to be maintained. "Furthermore, on a long journey,

14. http://www.thewall.nu/Architectuur (accessed at October 3, 2011).

Figure 4.4
The Wall, A2, Leidsche Rijn-Utrecht (Van Wijnen Waalwijk/VVKH Architects)
Source: Mark Kamphuis, MKFotografie.
Courtesy: Mark Kamphuis, MKFotografie, www.mkfotografie.nl.

it is useful to have visual uniformity all along the route" (OECD 1995: 157). In the Netherlands, both architects and road agencies started to question what they considered to be an increasingly unbalanced road appearance. Moreover, Dutch road authorities had every reason to believe that a more standardized approach would also be more cost-efficient.

Yet since these authorities saw that creating exclusive architectural designs was another important trend, they suggested *modular* designs for noise barriers as the *new* golden mean that would allow for both standardization and identification. This would then engender more visual continuity and architectural coherence along the road, and put an end to the "pattern card" of what barrier architects had to offer, the "salesmen architecture" along the highway, as one project leader had it (H.G. 2001: 34). This approach became known as the Thematic Approach Noise Reducing Provisions (Thematische Aanpak Geluidbeperkende Voorzieningen). Themes in barriers could tie in with the green walls appreciated by the "public on both sides of the screen." Planting could be integrated in modular systems for instance (Sloot 2002, Ministerie 1990: preface). Anything was better than a "muddled" (*verrommelend*) land, an "open air museum of less and more successful" barrier designs (Baauw 2008: 56). An example of the thematic approach was the recurrent use of a "migrating birds" motive in the design of barriers along the Dutch A2. (See figures 4.5 and 4.6.)

Figure 4.5
Screen "Migrating Birds," A2, Vinkeveen (Aletta van Aalst & Partners Architecten BV)
Source: http://www.aaenp.nl/project/1287654411/Vinkeveen–Geluidsscherm-A2, retrieved September 3, 2012.
Courtesy: Aletta van Aalst & Partners Architecten B.V., Amsterdam.

Figure 4.6
Yellow transparent screen, with "abstracted migrating birds," A2, Den Bosch (Ben van Berkel, UNStudio Architects)
Source: http://www.unstudio.com/projects/ringroad-a2-2, retrieved September 3, 2012.
Courtesy: UNStudio, Amsterdam.

The third trend was to articulate the local identity of the invisible place *behind* the barrier, a "glocalization" of the noise barrier that coevolved with the rise of city marketing. While an environment with noise barriers provided fewer and fewer signs that enabled people to orient themselves in space, art installations could stimulate their fantasy and create something "to really look at or talk about" (Meijer 1993: 53). In the United States, this tendency to flag the identity of towns and cities behind the scene *on* the wall became increasingly important at the beginning of the twenty-first century. It counterbalanced the loss of the driver's and residents' views, expressed the local areas' ecosystems, *and* functioned as community-specific visual icons. A concrete barrier along State Route 527 at Mill Creek (Washington), for instance, had designs "that featured maple and alder leaves, cedar branches, dragonflies, ladybugs, and animal tracks—all natural elements representative of Mill Creek's environmental heritage" (Sullivan 2003: 14). Since the speed was limited to 35 miles per hour, both pedestrians and motorists were thought to be able to appreciate the motifs. At Pima Freeway, Scottsdale (Arizona), however, the scale of the artwork on the residential side of the barrier was different from the motorists' side to account for differences in speed. The barrier expressed the local desert ecosystem, as well as the area's historical, cultural, and climatic diversities. "Elements included cacti, lizards, and other desert flora and fauna; mountains; and a Native American–inspired motif" (Sullivan 2003: 16). Again, the earlier strategy of compensating for people's loss of view of nature was combined with the new ideal of constructing a barrier

Figure 4.7
Pima Freeway, Scottsdale, Arizona, with representations of lizards, cacti, and Native American motifs. Design: The Path Most Traveled, by Carolyn Braaksma, Commissioned by the Scottsdale Public Art Program, Arizona.
Source: http://spa.contactdesigns.com/collection/pimafreeway.php, retrieved September 3, 2012.
Courtesy: Scottsdale Public Art Program, Scottsdale, Arizona.

as landmark or "icon of the community" (Billera, Parsons, and Hetrick 1997: 60). (See figure 4.7.)

In the Netherlands, architects did more than visualize local ecosystems. They expressed, for instance, the architectural history of towns in barriers, or the cities' traditional trade. As *CROW* admitted, municipalities didn't want to be concealed behind noise barriers any longer (CROW 2001: 95, 2007: 26). One barrier took the form of a round bastion or roundel (*rondeel*) after the architect had studied old maps (Ekkelboom 1999). Tiel, a town famous for its jam jars, boasted a barrier built with "area-characteristic" materials such as wood, referring to the orchards and glass, the jars (Baauw, Zwart, and Pontenagel 2004: 22). Whether or not these icons and landmarks were always perceived as such is beyond the scope of this chapter. Drivers definitely passed and still pass them in the blink of an eye.

Only one screen played with the image of sound itself: the Pipe Organ near Eindhoven, a "wood" of organ pipes. Its perforated pipes had highly

Figure 4.8
Tube screen, Highway A2, Eindhoven, the Netherlands (Van Campen Industries & VHP Architects)
Source: http://wurckweb.cronius.net/projecten/randweg-a2, retrieved September 3, 2012.
Courtesy: VHP/Royal HaskoningDHV.

noise-reducing and absorbing qualities and created an interesting visual spectacle (figure 4.8) (Anonymous 2008). But whatever the architectural identity of individual noise barrier, the bigger trend in road design was aimed at creating a contained and continuous flow of cars. And while this flow was increasingly impersonal, car radio afforded a more personal flow *within* the car.

NOTHING TO SEE, LOTS TO LISTEN TO: TRAFFIC RADIO AND AUDIO BOOKS

It was in late spring, 2011 that a famous Dutch columnist wrote an article expressing pity for people living near eight-lane highways. They were, he noted, exposed to the brutal and unbearable noise of traffic, day and night. "That's why the authorities have raised noise-reducing partitions. For miles and miles. In former days, you could watch the meadows, the pollarded willows, the cows, the farmhouses. Whether you enjoyed it or not, it was definitely rustic. Now you drive through a gray groove. But at least you can turn on the radio" (Montag 2011: 37).

Even though this columnist—a man in his eighties—wasn't very positive about what contemporary radio had to offer, he made something explicit that had not been articulated often before: a connection between driving the gray grooves of modern-day highways and listening to the gray grooves of recorded music. It was not that radio hadn't been associated with escaping the negative aspects of driving. As we have seen in the previous chapter, the car radio gradually transformed from a companion on dull drives into a mood regulator on overcrowded roads. As early as 1937, the BBC television program *Woman's Page* associated car radio with providing relief during long drives *and* traffic jams. Presenter and British novelist Ursula Bloom named radio her "most precious" car possession, a set "with which to while away long-waiting traffic jams or to relief you lovely of a long drive."[15] Yet up until the early 1960s, the link between car radio and traffic jams was rarely established. One reason may simply have been that only a small percentage of British cars had radios, though many people took radios with batteries along, considering it a "thrill" to drive in their "little cocoon" with "this music coming through."[16]

15. http://www.youtube.com/watch?v=xImnco66C_c (accessed August 25, 2011), at 4:30. We would like to thank Andy O'dwyer (BBC archives) for bringing this and other relevant BBC material to our notice.

16. *The Secret Life of the Motorway*, Part II, August 22, 2007, BBC DVD (courtesy Andy O'dwyer, BBC Archives).

Figure 4.9
Drive on wings: Philips all-transistor Autoradio (1964)
Source: Philips Company Archives Eindhoven, File 812.215, Advertisement material 1934–1960.
Courtesy: Philips Company Archives.

It was first in the 1960s and 1970s that the idea of the car radio as the driver's companion in coping with traffic jams became predominant. In 1971, the BBC television series *Tomorrow's World* featured an item about the use of "solid state radio" in the car. The anchorman heralded this type of car radio for its automatic tuning: it would make driving with car radio safer since it allowed drivers to keep their hands on their wheels while driving. But he also lauded the superb sound quality it afforded and the future option—already established abroad—of local radio stations breaching national radio programs with the latest information about the traffic situation. These new radio options would transform the car into a "magic carpet" that guaranteed a "smooth drive" on the road, the ultimate dream of every driver.[17] With an all-transistor car radio, a Philips ad had posited in 1964, one would "Drive on wings" (figure 4.9).

17. *Tomorrow's World*, item about "solid state radio," BBC Television Program broadcast on November 26, 1971 (courtesy Andy O'dwyer, BBC Archives).

Figure 4.10
Illustration accompanying an article on the introduction of the Autofahrer-Rundfunk-Informations System (ARI system), or Driver's Radio Information System, in New York (1983)
Source: Bosch Company Archives Stuttgart, File Bosch Archives, File 1601 623; Angus and Harrys 1983: 3.
Courtesy: Robert Bosch GmbH.

Indeed, providing traffic information for drivers and catering to the commuter with special programs had become an increasingly important function of radio. In the previous chapter, we mentioned several of these radio programs as well as the introduction of the Driver's Radio Information system, or ARI, in West Germany in the mid-1970s (figure 4.10). In the years prior to the adoption of ARI, a wide array of regional broadcasting stations using ultrashortwave, or FM radio, offered German drivers information on the need for snow chains in the mountains or the best checkpoint to cross the border, and music to "cheer up" their "Rush Hour."[18] It was not easy for drivers, however, to find the relevant broadcasting stations amid

18. Bosch Archives, File 1601 072, "Blaupunkt Autoradio" (1973): 2; File 1601 069, "Blaupunkt Autoradio" (1972): 2.

the myriad of FM stations at that time.[19] In the late 1960s, the German electro-technical industry examined the options for a special traffic radio net, but its costs and international radio frequency arrangements hampered the net's establishment. A few years later, however, a joint venture of car radio manufacturer Blaupunkt, the German automobile club ADAC (Allgemeiner Deutscher Automobil-Club), the Institute for Radio Technology (Institut für Rundfunktechnik), and German radio, postal, and traffic authorities ensured that a particular range of low-frequency radio bands would be assigned as index frequencies for traffic radio, which was officially adopted by the Federal Republic of Germany in 1974.[20]

The radio stations embedded in the ARI system acquired their traffic information from the German police and ADAC road service people, and broadcast this information through the frequencies assigned. Blaupunkt car radio then picked up the signals with help of a decoder, which had three versions. The most simple made sure that the car radio would distinguish between broadcasting stations offering traffic information and stations that did not, with a light informing the driver of the option to receive traffic information. A more sophisticated decoder filtered the radio stations that actually broadcast in the region the driver was crossing, and the most powerful one enabled the driver to receive traffic information even if the radio was turned off.[21] Unsurprisingly, the Blaupunkt pamphlets advertising the decoders came with photos of bad weather situations, accidents, crowded roads, traffic jams, and maps that showed that the ARI system covered nearly all of West Germany and had started to conquer Europe.[22] Indeed, the European Broadcasting Union soon advised the adoption of ARI in all of Western Europe, and Austria actually did so in 1976.[23] In turn, both the increasing use of FM radio and interest in stereo sound fostered the development of all kinds of automatic distortion suppression technologies for car radio.[24] In 1983, ARI even made it into the New York City area, where it drew on information offered by "Shadow Traffic, a company

19. Bosch Archives, File 1601 087, "Blaupunkt Autoradio-Autostereo-Autofunk '77/78" [1977]: 5.

20. Bosch Archives, File 1601 087, "Blaupunkt Autofahrer-Rundfunk-Informations-System" [1977]: 7 (with map on p. 5); File 1601 623, "Blaupunkt Moderne Verkehrslenkung durch den Rundfunk" (1977): 5–6.

21. Bosch Archives, File 1601 072, pamphlet "Blaupunkt Konstanz Dortmund" [1974]: 4.

22. See the references in the previous three notes, and Bosch Archives, File 1601 081, pamphlet "Das Blaupunkt System" (1976): 8.

23. Bosch Archives, File 1601 623, "Blaupunkt Moderne Verkehrslenkung durch den Rundfunk" (1977): 7.

24. Bosch Archives, File 1601 082, "Die automatisch Nachrichten-Zentrale," "Sonderdruck aus 'Auto-Zeitung' Nr. 12 bis 15" (1976): 34; "ASU Automatische FM-storingsonderdrukker."

that provides traffic reports gleaned from a network of airplanes, helicopters, spotter cars, direct phone lines to the tunnel and bridge authorities and police departments, and . . . 'regular citizen-type people,' who take the same routes each day to their jobs and call in two-way radio reports to Shadow."[25] The magazine article announcing this news intriguingly reinvented the trope of the magic carpet again, now in an invisible version, as a retouched photo illustration shows.

In the UK, traffic information was gathered with help of "the flying eye" in the 1980s, only to be replaced by networked cameras and computerized control centers in the twenty-first century—as the excellent 2007 BBC television documentary *The Secret Life of the Motorway* shows. At the end of the 1980s, Dutch motorists said they listened to radio traffic information very frequently: 90 percent of a representative sample used traffic information, 44 percent even a few times a day (Akerboom 1988: 46). Most of them used it quite regularly to decide on which route to take, and actually asked for more frequent, trustworthy, and timely information that would warn the driver and give details about the length of traffic jams, alternative routes, or ways of driving in foggy weather. The report that documented their wish list had high hopes for digital technologies in this respect (Akerboom 1988: 36–38). Clearly, car radio and the desire to drive "in control" had almost become Siamese twins. Traffic information has also increasingly become available through Dynamic Route Information Panels (DRIPS) and GPS-based visual in-car navigation systems. Yet for safety reasons, radio—like the Traffic Message Channel of the Radio Data System installed in the Netherlands in 1998—is still considered the principal way to inform the driver about the situation on the road (H.G. 1999, 2000a, 2000b). Today, as British radio disk jockey Tony Blackburn explains in *The Secret Life of the Motorway*, there are "two places where you have a really intimate relation with the listener. First of all there is that relationship with the listener who is in the car, because there is a *captive* audience. The second one is anybody in prison!"[26]

The radio man flashes a joking smile after this remark. Yet what he says is fully in line with the idea of car radio as mood regulator, magic carpet, and music groove added to the gutter groove. Car radio and other audio equipment literally and metaphorically provide alternatives to being "caught" on the road: caught by the road rage of fellow drivers, caught in traffic jams

25. Bosch Archives, File 1601 623, Angus and Harry 1983: 3.
26. *The Secret Life of the Motorway*, Part II, August 22, 2007, BBC DVD (courtesy Andy O'dwyer, BBC Archives). Italics added by the authors, yet Tony Blackburn underlines it verbally.

or speed control, caught between the oppressing walls of noise barriers—caught in the "laws" of a commuter's existence. While the predominance of traffic jams, noise barriers, and matrix panels with speed limits may vary from country to country and from region to region, no driver will experience a jam-barrier-panel-free driving life.

One episode of *The Secret Life of the Motorway* centers on a couple who seem to be in their late thirties. They live in Warrington, a British "commuters' heaven." A third of its working population heads out every morning. So do both partners, each in his or her own car. One would expect some sense of misery attached to their driving lives, which is actually not the case. Rather, they both enjoy the drives. She likes "the time on her own," and hates it when other people use her car. She "hides chocolate in the glove compartment" that her daughter then finds. He has recently "developed a bit of a weakness for teenage dance music" while driving. In addition, he loves Italian food and Italian cars and has started taking an Italian-language course. Watching the documentary, we hear Italian opera and the Italian course sentences on an audio compact disk or MP3 player. It is quite amusing for other drivers, the man adds, to see him talking behind the wheel and "gesticulating as if I am completely insane, but who cares, it's my car, it's my space." Meanwhile, his wife has been "writing stories in her head." Her favorite story, one she started making up after a bad day at work when she did not want to go home, is about someone who keeps driving endlessly.[27]

While the camera follows her as she heads for her destination, we see noise barriers in the background. She uses the relative silence and privacy of the car to think up stories about escape. He uses it to fill its sonic space with teenager dance music or Italian voices. To both halves of the couple, the car provides relief and diversion, and "sound" clearly contributes to this feeling. It may thus not be a coincidence that given the ever-expanding automobility in the West, the popularity of audiobooks and downloads is on the rise. In 2010, the Association of American Publishers reported a compound annual growth rate in audiobook sales of 4.3 percent over 2001–2008, followed by a decrease in sales in 2009, and then a rapid increase again in the first half of 2010 (over 14.7 percent for audiobook sales and 32.5 percent for downloads).[28] Two years earlier, the Audio Publishers Association had also reported growth up until 2008.[29]

27. *The Secret Life of the Motorway*, Part II, August 22, 2007, BBC DVD (courtesy Andy O'dwyer, BBC Archives).

28. Industry Statistics 2009: http://www.publishers.org/main/IndustryStats/indstats_02.htm (accessed August 26, 2011); http://audiobookupdate.com/reasons -audio-book-sales-increasing-2010/ (accessed February 7, 2011), 1.

29. "More Americans Are All Ears to Audiobooks," press release Audio Publishers Association (APA) September 15, 2008, 1. Between 2009 and 2010, the number of

We cannot actually prove a causal connection between the expanding use of the car on the one hand and of audiobook sales and downloads on the other. We have to bear in mind that the production of audiobooks on compact disk and MP3 files would not have been possible without the ongoing digital revolution, and that the relatively low prices of recent audiobooks (due to reduced packaging, storing, and transportation costs) play a role.[30] Neither should we ignore the long-standing tradition of spoken recordings of literary works, going back as far as the 1890s, book-length "talking books" on the gramophones of the 1930s, and the audiobooks on magnetic audiotapes, compact cassettes, and LPs of the 1950s, 1960s, and beyond (Camlot 2003; Rubery 2008, 2011). Yet we *do* know that audiobooks are most widely used in cars. As a consumer survey by the Audio Publishers Association showed, audiobook listeners "are most likely to be listening in their cars."

> In addition to being the most frequent reason for why audio book consumers begin listening, entertainment for a long drive is also the most frequent reason (40 percent) why they continue listening. The opportunity to listen while engaging in another activity (23 percent) and while commuting (18 percent) are also common reasons, according to the Consumer Survey.[31]

Similarly, we cannot claim that noise barriers caused the rise of the audiobook. But while drivers consider noise barriers anything but entertaining, audiobooks enliven their long drives. Once again, drivers have found a way to use sound technologies in order to create sonic relief in their cars.

A REENCHANTED SPACE

Audio equipment and audiobooks help drivers to create an auditory ambiance that compensates for their dramatic loss of views of the landscape-garden along the road and their reduced control of driving conditions. These artifacts also sonically furnish the highly artificial space of

audiobooks sold in Germany rose from 16.1 to 16.4 million, while the number of downloads of audiobooks in 2010 was 2.7 million, 18 percent higher than in 2009. See http://www.boersenverein.de/de/158446/Hoerbuch/158293 (accessed October 6, 2011).

30. http://audiobookupdate.com/reasons-audio-book-sales-increasing-2010/ (accessed February 7, 2011), 1.

31. "More American Are All Ears to Audiobooks," press release Audio Publishers Association September 15, 2008, 3.

the car, the reenchanted space Rosalind Williams has considered a response to the many man-made environments that degraded nature. While drivers associate the road fitted out with noise barriers with an ugly, overwhelming, and nearly inescapable underground, and lament the fall of the cinematic drive, the car itself has become an important refuge. It has transformed into an enclosed interior for auditory privacy and sonic control, and a magic carpet when amplified with radio traffic information.

The societal concern about the exterior noise passenger cars and other forms of road transport generated has given rise to the noise barrier. When governments started measuring the average levels of noise immission their citizens had to put up with and consequently set limits to these levels in the late 1970s, the noise barrier was considered a relatively cheap and easy way to reduce the noise transmitted to residents living close to highways. It did not take long before authorities noted the visual problems connected to noise screens, and drivers started to articulate what is was that didn't love these walls: the oppressing effect, the experience of deepness, of being dragged along, and the hardly voluntary seclusion from nature. Denial design was the initial response: greenery-camouflaged barriers, transparent barriers, and barriers that projected onto the wall the landscape that was hidden behind it. Yet from the late 1980s and early 1990s onward, the towns and cities behind the wall began to make themselves heard and wanted to speak through the barrier to become once again identifiable to the passing driver. Together with an opposing trend, the introduction of standardization, modular design, and trajectory themes, this led to the glocalization of the noise barrier. What's more, this glocalized barrier had to move drivers by moving *with* them.

As we acknowledged, the importance of radio, other audio equipment, and innovative listening practices in the car is not solely due to the situation on the road, and certainly not only to the rise of noise barriers. A recent study among employees in Sweden of their satisfaction with the office environment has shown that today's open-plan offices have engendered widespread displeasure with noise and the lack of auditory privacy in workplaces (Bodin Danielsson and Bodin 2009). The need for auditory privacy and sonic relief in the car may thus be a response to societal changes that go beyond traffic control and noise barriers on the highway. Clearly, motorists are among the most captive radio audiences and audiobook listeners. Listening to radio and other audio devices in the car at least *enables* drivers to adapt to their diminished control over *how* to drive and *what* to look at while driving, notably in a country like the Netherlands, where roadside noise barriers and fellow drivers abound.

That drivers feel free to listen to their audio equipment instead of their car's engine is the result of their being confident about their cars' functioning as well as the expert ears and other diagnostic skills of their mechanics—the story of the previous chapter. Drivers no longer need to constantly check whether their car is running properly. And when they do hear something unfamiliar, they see their mechanic or dial into a radio program featuring automotive experts. Yet if drivers no longer need to listen to their car, why do today's car manufacturers find it so important that everything makes the sound it is supposed to make? The answer is in the next chapter.

CHAPTER 5

Selling Sound

Sensory Marketing in the Automotive Industry

OVER DINNER

Up until the late 1990s, the evening dinners of ISO Working Group 42 dealing with the measuring of noise emission from road vehicles had always been relaxed. Even when its members had been involved in heated discussions during the day, the dinner would clear the tensions built up earlier. The participants would shift their topic of conversation from sound levels, algorithms, and the position of microphones to family, football, and the latest gadgets. After years of compliance with this tacit code, however, the atmosphere would unexpectedly remain tense, even during dinner. On the agenda was the decision about what to measure *exactly* in car sound. Should all drive-train sounds emitted by cars be taken together in one measurement—the sounds of *all* moving components of the car such as the engine, clutch, transmission, driveshaft, axles, wheels, and tires—or should measurements be split up, for instance into readings of power-train sounds, like those of the engine, and readings of tire noise? Representatives of the automotive industry wanted to be as inclusive as possible, while several of those speaking on behalf of agencies involved in developing noise abatement strategies promoted measurements that focused on particular sources of sound, one by one. But why would experts from the automotive industry be more inclusive than the ones with governmental links? And why did the clashes during the days become so fierce that their emotional effect could not be easily smoothed out over dinner?

In 2011, American car manufacturer Chevrolet posted a short movie on YouTube.[1] The film shows an anechoic test bay at Milford Proving Ground. A Chevrolet Cruze Solid is on trial, or more precisely, its *sounds* are on trial. Two staff members start their high-precision work: "Testing, one, two, three." The film begins rather quietly with the repeated slamming shut of a car door. "Solid" is the assessment of the testers. These sounds soon transform into an exciting rhythmic pattern of slapping car doors followed by and combined with the sounds of the Chevrolet's wipers, tires, horn, trunk, safety belts, windows, wheels, switching board, and glove compartment. The result is a two-minute music video that would perfectly suit a techno dance party.

At first sight, these two vignettes seem unrelated. Yet an initial answer to the questions raised by the first account can actually be found in the second. Chevrolet's short film fits into a series of car sound television commercials launched in the 1990s. We mentioned these commercials in the opening chapter. Advertising car sound was anything but new—as we have seen in previous chapters—but television spots featuring car sound and nothing else besides *were* a novelty. These commercials not only indicated car manufacturers' growing interest in sound design, but also voiced their wish to attract specific groups of consumers through "target sounds" expressing a particular car's brand identity. These trends offer the key to understanding the troubled atmosphere in the ISO working group. The stakes behind the clash about measuring sound levels concerned the freedom to choose *which* sounds would have to be subdued—a freedom treasured by the car manufacturers—versus the obligation to focus on particular sources of sound, such as the engine, as was the wish of those fostering noise abatement policies.

This chapter explains how the car manufacturers' aim to remain in charge of trading the silence of particular components for the noises of others has been directly related to their intensified interest in sound design and marketing. We will first show how much car manufacturers and other stakeholders have invested in sound in the past decades. We will then unravel how the increasingly tense atmosphere at ISO dinners reflected both the character of the discussions in Working Group 42 and the high stakes behind the scenes of standardization. Since we had the unique opportunity to study the documents that circulated within this working group and to interview several of its key members,[2] we are able to explicate

1. http://www.youtube.com/watch?v=Oze4fmuOVxo&feature=player_embedded# (accessed on December 20, 2011).

2. For an overview of the interviewees from ISO Working Group 42, see "Archives, Journals, and Interviewees."

how something as seemingly innocent as humming engines had become a bone of contention between engineers fighting for the success of automotive industries and engineers fostering environmental policies.

Issues of standardization made car sound a highly complicated area of design. As we will show in a subsequent section, however, sound design happened to be even more complex: the automotive industry had a hard time finding out how consumers perceived and evaluated car sound, and which target sounds they exactly wanted. Even though the 2011 Chevrolet film zoomed in on laboratory testing by experts, the interest in target sounds in the automotive industry went hand in hand with the introduction of real-life sound-testing by everyday drivers and the novel use of qualitative methods for capturing their evaluations. The latter shift was a highly contested moment in the search for "the holy grail of psycho-acoustics" (Fidell 1996) and marked the beginning of what would develop into a new tradition of testability. We acquired in-depth insights from the car-manufacturing industry and acoustic consultancy world into how the people working in these areas tried to tackle the problem of understanding the sensory experiences of drivers by studying one episode in particular. This episode is the collaboration between the automotive industry and research partners in the EU-funded project OBELICS—a project focusing on car sound perception and design, and coordinated by the acoustic consulting firm HEAD Acoustics GmbH. This "gatekeeper" opened up the doors to the automotive industry and allowed us to interview staff members from the acoustical engineering departments of Ford, Opel, Renault, BMW, and Porsche, most of which had been formerly involved in OBELICS. The interviewees not affiliated with OBELICS were all experts working on testing, design, and marketing of sound and on noise control.[3] In addition, our analysis relies on field notes from the observation of testing sites, documents produced in the context of OBELICS, and publications by acousticians and the automotive industry.

But what prompted the heightened significance of sound design in the 1990s? And what did it mean for the emergence of acoustic cocooning, for making the driver feel "sound and safe"? We will answer these and previous questions by using sociological theories on the rise of the experience society and emotional capitalism, as well as theories on standardization and testing from science and technology studies. Together, these strands in

3. For an overview of the interviews with employees from acoustic consultancy firms and car-manufacturing companies, participants in sound design research projects, and EU officials, see "Archives, Journals, and Interviewees." This overview lists information on the interviewers as well as the location of the interviews.

scholarly work will help us understand why many car manufacturers have become so keen on offering every type of consumer their own *choice* of car sound—like the choice of mobile phone ringtones—thus instructing drivers once again on a new form of acoustic cocooning.

COMING TO YOUR SENSES: COMPOSING THE SOUND OF CARS

We propose to watch another YouTube movie, a television commercial widely broadcast in 2010–11. It advertises the Peugeot 508, a hybrid. A series of ultrashort shots show scenes in the life of a handsome businessman. We observe him putting a cell phone in a docking station, having coffee, talking on the phone, checking his Blackberry, drawing the curtains of what seems to be a hotel room, taking a shower, answering the phone again, waiting for files to upload, having lunch, shaking hands, attending a meeting, reading papers, taking notes, and, finally, loosening his necktie. All accompanying sounds are loud. Then, suddenly, we see him closing the door of his car, and tranquility enters the scene—he smiles and turns his car onto the road, his right hand reaching out for the car radio.[4]

The Peugeot commercial is a recent example in a long-standing series of television spots on car sound. Remarkably, this commercial not only stresses the peace and quiet to be found in the car but also refers to the sound it offers: that of car radio. In the 1990s, most of the car sound commercials merely focused on the car's silent interior. A 1999 Toyota Avensis commercial, for instance, shows a university professor type in a library having a hard time focusing on his work, distracted as he is by the noises made by those around him: someone is tapping a pen, another person crumbles up a piece of paper, a mobile phone is ringing, a Walkman sounding. He flees from the library, heading straight for his car. For a short while we hear the sounds of the city. Yet the moment he pulls the car door shut he begins to unwind: quiet at last.[5] A commercial for Mercedes-Benz, from 1990, has a very similar plot. A businessman arrives at Tunis airport: it is hot and smelly and the aircraft engines roar. The airport itself is immensely crowded. The man has rented a car, but it takes some time to actually find it amid the visual and sonic chaos of the North African city. As soon as he

4. See http://www.youtube.com/watch?v=qIk9Rce8ELA (accessed January 16, 2012).
5. *Zelfs de stilte is standaard* (1999), commercial Toyota Avensis, videotape. Talmon: AV Communicatie.

spots his rental Mercedes-Benz, however, he brightens up. Once inside the car, he heaves a sigh of relief.[6]

A sigh of relief also features prominently in a Volkswagen Passat commercial from the late 1990s. A concert by the Blind Brothers Three from New Jersey ends with a swinging song. The audience responds enthusiastically: they yell, applaud, and whistle. The Blind Brothers leave the venue by cab. In the cab, all is quiet. "Lord," one of the musicians mumbles with relief, "this is moonlight." The three evaluate the concert and nearly start quarreling about a wrong note. These guys notice everything, that's for sure. Then one of them opens the window. We hear the wind blowing and the traffic roaring. His colleague is surprised to find that they are actually moving. "Brother, what kind of limousine is this?" "It's a Volkswagen, Sir," the cab driver replies. The Blind Brothers roar in disbelief: "Yeah, and we are the Pointer Sisters!" Yet *we* see it *is* a Volkswagen, a car with the sound of a luxurious limousine—at least that is the commercial's message.[7]

Some of the more recent commercials depart from this tranquility plot by not only featuring the car's silence, but also its distinctive sounds, as the Peugeot 508 movie did with car radio. In one of these more recent commercials, we see a man driving a small yellow car and parking it somewhere in the city. The man is singing, whistling, and grumbling all the sounds he would like his car to make: the sounds of a remote control, sliding window, safety belt warning signal, car radio, horn, navigation voice, parking control, and alarm. At the end of the commercial, we see the car he fancies: a Citroen C3, "loaded with specs."[8] In a Porsche commercial, a man parks his Cayenne on a lonely hill with a magnificent view of the city. He starts testing the sounds of his car's engine, and Porsches throughout the city, including those produced in earlier decades, seem to respond by echoing these sounds. "The bloodlines are unmistakable," the voiceover claims.[9] Our final example is one of our favorites. A two-minute movie by Honda features an a cappella choir in an underground garage that sings and hums the sounds that are meant to express "what Honda feels like." With unprecedented creativity the choir mimics the sounds of a Civic moving on asphalt and cobblestones, taking ramps, accelerating, slowing down, turning sharply, with

6. *Business Travel* (1990), commercial Mercedes-Benz, videotape. Stuttgart: DaimlerChrysler AG.

7. Blind Brothers Three (1999), commercial Volkswagen Passat, videotape. Almere: Team Players.

8. See http://www.youtube.com/watch?v=adbBm0gK2og&feature=related (accessed January 24, 2012).

9. See http://www.autoblog.com/2008/01/09/video-porsche-cayenne-gts-bloodlines-commercial/ (accessed January 24, 2012).

its windows opening and closing, its windshield wipers moving in an arc, and the wind gliding along its body.[10]

This shift from mere silence to silence *and* sound in advertising cars is no coincidence: it expresses automakers' strengthened confidence in their ability to enhance the overall quality of driving with respect to its various auditory dimensions. Indeed, manufacturers have invested considerable time and money, notably since the 1990s, to make sure that almost *everything* in the car—from engines, the locking of car doors, to switches, warning signals, direction indicators, windshield wipers, and the crackle of the leather upholstery—comes with the *right* sound. Three simple figures demonstrate just how much today's automotive industries invest in their product's sonic characteristics: Opel has 80 employees working on sound,[11] BMW employs over 150 acoustic engineers, and Ford has an acoustical department of about 200 employees (Jackson 2003: 106; van de Weijer 2007: 9; RH: 20).[12] Other companies, like Renault and Fiat, collaborated in projects on car sound design funded by the European Union. In the 1990s, one such project was OBELICS. We will discuss this project below.

It has not only been the search for target sounds, however, that stirred exterior and interior car sound design. In the 1960s and 1970s, national governments and European organizations imposed increasingly stringent limits on the maximum noise emission of cars (Sandberg 2001; DD: 1–2; Bijsterveld 2008). Between 1970 and 1995, the European maximum for passenger cars decreased from 82 dB(A) to 74 dB(A) (EL and VM: 6). As a result, automobile makers were forced to further limit car noise. Moreover, when it appeared that cars with very low levels of sound, such as hybrid and electric cars, were dangerous because pedestrians did not hear them approaching, car manufacturers started experimenting with the idea of having sounds in electric cars uploaded (Anonymous 2006: 77; van Kleef 2002). In 2010, the United Nations Economic Commission for Europe (UNECE) even worked on "a set of minimum noise standards designed to keep quiet vehicles from becoming a safety risk to pedestrians, notably the blind."[13]

10. See http://www.youtube.com/watch?v=gjyWP2LfbyQ (accessed May 15, 2012).

11. http://www.youtube.com/watch?v=MJesIniGVjM (accessed January 12, 2012).

12. In references to interviews, we give the initials of the interviewee and the page of the interview transcript. For the names behind the acronyms, see "Archives, Journals, and Interviewees."

13. http://www.youtube.com/watch?v=j5M4UD9AK8Y and http://wn.com/exact/UNECE (accessed January 24, 2012). See also http://www.greencarreports.com/news/1056877_has-u-n-doomed-quiet-electric-cars-to-a-life-of-noise-making (accessed January 24, 2012). At the end of 2011, however, the European Commission decided *not* to create mandatory minimum noise levels; see http://www.europeanvoice.com/article/2011/december/commission-proposes-vehicle-noise-limits/72924.aspx (accessed January 24, 2012).

Yet in the 1970s and after, *limiting* exterior car noise became the dominant trend. The number of articles published in five international automotive and acoustic journals and periodicals on how to reduce such noise increased from seven in the 1950s, to eleven in the 1960s, to twenty in the 1970s.[14] In the 1950s, engineers studying exterior car noise still mainly approached it as a safety issue—noise filled drivers' ears, took away their nervous energy, and created fatigue, and was therefore potentially dangerous during long drives—an attitude many authors writing on noise had taken in the 1920s and 1930s (see chapter 2).[15] This was linked to the fact that commercial vehicles, notably trucks fitted with diesel engines, acquired most attention: the exterior noise audible within these truck cabins was considered to deteriorate the working conditions of the drivers. Some of the noise control innovations for commercial vehicles trickled down to passenger car design. Among them were exhaust mufflers, the shielding of inlet valves, and a more efficient mounting of engines through new designs of the bolts holding the engine to the frame. At times, engine innovation hampered noise control. Compact engines with higher horsepower generated more heat, which in turn required the use of cooling fans, a major source of noise. Raising compression in the combustion engine also resulted in higher noise levels. Improvements in fuel efficiency, weight, aerodynamics, friction, vibration, and wearing, conversely, led to an unplanned reduction of noise.[16]

By the 1970s, car noise control had lost its relative innocence. It had turned into an environmental policy issue the automotive industry simply *had* to deal with, even under increasingly difficult circumstances. Small, low-performance vehicles, for instance—the heroes of an energy-conscious age—required "higher engine speeds and greater throttle openings to produce a given acceleration speed," which did not make such cars more quiet (Marks, Fischer, and Stewart 1974: 34). The interventions chosen were geared to a smoother operation of the car or to sound insulation, for instance through the use of sound-dampening surfaces of engine walls.[17] Engineers had their hands full when it came to reducing the low-frequency

14. These five journals and periodicals are *Journal of the Society of Automotive Engineers*, *Journal of Sound & Vibration*, *Noise Control Engineering*, *FISITA Proceedings*, and the *Automotive Engineer*. In 2008, the Eindhoven University of Technology students Gerald Bachler, Jorrit Bakker, Sander van Cuijck, Sylvia Laurensse, and Assal Razavifar (MA Technology and Policy) finished a quantitative and qualitative analysis of these journals under supervision of Gijs Mom and Stefan Krebs: "Car Noise Control/Results: Research about Car Noise Abatement (1950–1980)," 3 and 73.

15. "Car Noise Control/Results," 24.

16. "Car Noise Control/Results," 35–41.

17. "Car Noise Control/Results," 40.

Figure 5.1
Mercedes S test bench (2005)
Source: Hoed 2005.
Courtesy: Mercedes-Benz Nederland BV.

sound in cars (Stockfelt 1994). There were many such sounds: the rustle of the wind, engine vibrations, the sound of tires touching the road surface, the hydraulic system of power steering, and the whining of the fuel pump. Noise, vibration, and harshness (NVH) engineers tried to tackle these sources of noise by, for instance, applying acoustic glass or introducing innovative profiles for tires (Dittrich 2001; Kouwenhoven 2002; Steketee 2006). Creating stiff, multilayered and asymmetrically ribbed floor plates, hermetically closing off cable entries, and filling hollow spaces with foam composites—a technique developed by the plastics industry in the 1970s[18]—were other options (den Hoed 2005). (See figure 5.1.)

By carefully distinguishing between power-train noise, road noise, and wind noise, by discriminating structure-borne from airborne sound, and by eliminating noisy features, engineers gradually created room for the design of particular target sounds (Govindswamy et al. 2004). Time and again, however, they would find that removing one noise rendered another one audible (Schick 1994; Freimann 1993; NC: 20). So, even though car manufacturers had been interested in interior car sound since the early 1920s, it was the obligatory reduction of *exterior* noise that brought formerly masked *interior* sounds into the limelight again (WK: 8; BL: 11;

18. "Car Noise Control/Results, 41.

RoS1: 11; SP: 13). Indeed, the rise of quieter forms of asphalt, such as "whisper asphalt," tested in the early 1970s and increasingly predominant on Dutch and Flemish highways from the mid-1990s onward, kept changing the balance of sounds audible in the car.[19]

As an Opel employee claimed in a short documentary on sound design, engines might be compared to pianos. In doing sound design, he said, one needed to be "a bit of an artist," in addition to being a "mechanical engineer or a physicist."[20] All three laureates in a 2011 German contest for the best series of sounds designed for an electric car were in fact composers.[21] Moreover, the first car make with Vehicle Sound for Pedestrians (VSP), the Japanese Infiniti M35h, reportedly sounded like a mix of vacuum cleaners and Star Trek—a blend attributed to the composers of the score (Oosterbaan 2010: 12).[22] As we will see, however, even combining the qualities of an engineer and an artist did not suffice in designing consumer-specific car sounds. The listening habits and emotions of consumers had to be understood, and ISO standards had to be known and taken into account. Which standards exactly?

LEEWAY FOR CAR SOUND DESIGN: BEHIND THE SCENES OF INTERNATIONAL STANDARDIZATION

We are hardly aware of it, but our day-to-day functioning, especially when it involves technology, is deeply affected by standardization activities, such as those of the Geneva-based International Standardization Organization (ISO). From the plugs that feed us with electricity and the colors of the toys of our children to the containers in which we ship consumer products from one part of the world to another: ever since its establishment in 1947, ISO has often had a say in it.

Surprisingly, however, Geneva has no archive that can inform us on how these standards have developed. Of course, one can check the standards and the different versions of these standards published over time. Yet what happens in the ISO technical committees, subcommittees, and working groups

19. See http://nl.wikipedia.org/wiki/Zeer_Open_Asfaltbeton (accessed August 28, 2012).

20. See http://www.youtube.com/watch?v=MJesIniGVjM (accessed January 12, 2012).

21. See http://www.ika.rwth-aachen.de/veranstaltung/2011/12-14/index.php (accessed December 20, 2011).

22. This system is switched off as soon as the car moves faster than 30 kilometers per hour.

is a mystery to outsiders and can normally only be known by those who are members. Most of these members are permanent, representing national standardization institutes such as the British Standards Institute. Such permanent members have, unlike correspondent and subscriber members, a right to vote in the standardization process. It has been through such permanent members and the convener, or chair, of Working Group 42 that we acquired access to ISO files.[23] These people store their files in their offices at work or at home in order to keep track of the complicated and often time-consuming process in which a proposal for a new work item proceeds from a preparatory phase, to a Final Draft International Standard (FDIS), to approval—if everything progresses smoothly.

When ISO Working Group 42 was created, in 1993, the standardization of measuring car noise emission already had a history of over fifty years. As soon as national governments introduced numerical limits for car noise, tests were needed for measuring it, either *stationary tests* that helped catch offenders on the spot, or *drive-by* tests for type approval of cars.[24] The German Road Traffic Act of 1937, for instance, mentioned a drive-by test that measured the noise of a single car with wide-open throttle at a speed of 40 kmh from a distance of 7 meters. Taking the test was actually rather tricky as the driver had to operate at full throttle and use the brakes at the same time to stay below the speed limit (Stadie 1954). In contrast, France only included a measuring technique in its traffic noise legislation by the mid-1950s, while opting for a measuring distance of 10 meters and a drive-by speed of 40–60 kmh pending on the vehicle category being tested (Thiébault 1963). Such drive-by tests were highly important to the automotive industry, notably after the formation of the European Economic Community (EEC) in 1957. The gradual abolishing of internal EEC automobile import duties fostered intracommunity trade tremendously and stimulated the circulation of car makes from different countries across the Common Market (Krebs 2012b: 28–29). This development enhanced the need for harmonization of noise-measuring techniques, however. It was important to ensure that differences in measuring techniques would not thwart the type approval of cars across national boundaries.

It was hence no coincidence that ISO Technical Committee for Acoustics (TC 43) was asked to develop an international car noise measuring

23. We would like to express our gratitude to Foort de Roo (Dutch former member of ISO Working Group 42) and Leif Nielsen (Danish secretary of Working Group 42) for allowing us to examine these archives.

24. Cars that pass such type approval tests can be sold on international markets.

standard, which caused TC 43 to set up a working group in 1958. As Stefan Krebs has shown, ISO Regulation 362, resulting from this working group's activities in 1964, nicely integrated bits and pieces from earlier national measuring tests. The required size of the test site, weather conditions, and measuring distance evolved from the 1937 German Road Traffic Act, details about driving conditions came from the French, and the choice for measuring sound levels in terms of dB(A) could be traced to two influential British members of TC 43 (Krebs 2012b: 31–32). This is fully in line with what historians and sociologists of standardization have argued previously: behind the language of universally valid knowledge often persist local traditions in research and testing. National industries, research institutes, and standardization bodies tried to reduce the costs of changing their measurement habits or aimed to enhance the prestige of locally produced test instruments by transferring what was familiar to them into international standards and harmonized regulations (Schaffer 1999, 2000; Pfetch 2008).

The number of countries actually employing ISO/R 362 rapidly expanded when the Working Party on the Construction of Vehicles (WP 29) of the United Nations Economic Commission for Europe (UNECE) adopted it. In 1970, the EEC included it in its Council Directive 70/157, which issued a maximum permissible noise level of 82 dB(A) for passenger cars.[25] Two years later the Organization for Economic Co-Operation and Development (OECD) also recommended noise limits to be measured with the help of ISO test procedures. The United States was not among these countries, however. The US Noise Control Act (NCA) of 1972 enabled the Environmental Protection Agency to impose maximum noise levels for commercial vehicles, yet not for passenger cars—most states had their own regulations for cars. Moreover, these regulations referred to a test procedure developed by the Society of Automobile Engineers (SAE) rather than ISO—a major difference being a measuring distance of 15 instead of 7.5 meters (Krebs 2012b: 37). Ford engineers clarified why they were not too keen to adopt the ISO standard. Once they tested a high-power sedan at "European distance," it emitted 88.3 dB(A), which was 6.3 dB(A) above the EEC limit. With the American procedure the result was 82.3 dB(A), which complied with most US regulations (Vargovick 1972; Wesler 1974).

But even within ISO, Regulation 362 did not remain uncontested. After an initial revision in 1981, Working Group 42 was asked to create

25. See http://eur-lex.europa.eu/LexUriServ/LexUriServ.do?uri=DD:I:1970_I:31970L0157:EN:PDF, 112 (accessed August 27, 2012).

a measuring method that was both more "representative" of urban driving conditions and more "reproducible" than its predecessors. As to representativeness, the new standard had to catch up with novelties in car technology such as a shift from five to six-speed gearboxes and new sorts of tires.[26] In 1993, the Swedish traffic noise expert Ulf Sandberg noted that tire/road noise had probably predominated in producing passenger car noise since the mid-1970s. Yet "new types of tires, especially with increased width" had made things even worse (Sandberg 1993: 49). In an interview, one of the former members of Working Group 42 endorsed this opinion and pointed out that the tires had not only become much wider, but increased in diameter, enhancing the levels of noise the tires produced. This had raised questions about which sources had to be taken into account when measuring noise. For quite some time, the exhaust had been considered the main source of car noise, and ISO/R 362 had been designed in accordance with this assumption, requiring test driving "with a relatively high number of revolutions per minute, and full-throttle." The idea had been that such test conditions would foster improvements in exhaust design. In the meantime, however, tires had become the new big problem (EG: 9). Moreover, an "acceleration rate at wide open throttle"' was not typical for urban driving conditions.[27]

Another new reality was that the European Union had gradually given environmental noise higher priority on the policy agenda, with a reduction of the upper exterior noise limit of passenger cars to 74 dB(A) in 1995, and the publication of the *Green Paper on Future Noise Policy* (1996) as important stepping stones. As an advisor of the Dutch Ministry of Infrastructure claimed about the work within Working Group 42: "It used to be one-way traffic, because the [automotive] industry was simply completely in charge. Over the past few years, this has changed somewhat. The [European] Commission is held accountable for progress in the sound field, so the paradise is sort of over" (BK: 11).

Even though paradise was "sort of" over for the automotive industry, or better still *because* paradise seemed to be over, the composition of Working Group 42 shifted from a balanced mix of employees from applied research institutes (either independent or government funded), testing

26. Personal archives Foort de Roo, File ISO TC 43/SC1/WG 42, unnumbered documents 1993–2001, Summary of the First Meeting, Paris, November 2–3, 1993, 1; Summary of the Second Meeting (by R. F. Schumacher), Seattle, June 13–14, 1994 : 1.

27. Personal archives Foort de Roo, File ISO TC 43/SC1/WG 42, numbered documents D1–D84, D65(I), email "Draft Format" from R. F. Schumacher to member of WG 42, July 8, 1997: 1.

agencies, and car-manufacturing industries to a mix in which people from the automotive industry predominated.[28] Among the first category were, for instance, the Swedish Road and Traffic Research Institute and Malcolm Hunt Associates, a New Zealand Environmental Noise Consultancy. Car manufacturers involved were General Motors, Fiat, Nissan, BMW, Porsche, Toyota, and PSA (Peugeot-Citroen). Clearly, there was something at stake. As we explained earlier, permanent members of ISO committees and working groups are sent by and represent national standardization bodies. Yet these national bodies are free to choose whether they send experts linked to relevant industries or experts normally working for research institutes or consultancy firms. In case of WG 42, the experts came from countries as divergent as the United States, Australia, Germany, Italy, France, Japan, the Netherlands, Norway, and Sweden.

Unsurprisingly, countries with large and important automotive industries like Germany and the United States sent experts from manufacturing companies, while countries without such industries, such as the Netherlands, opted for experts affiliated with consultancies and research institutes. As the original US convener of Working Group 42 explained:

> When one or two ISO procedures were promulgated that had a direct impact on the products being sold worldwide, including Europe, it [be]came obvious [that] the US manufacturers needed to be aware of these procedures earlier in the process and if possible have a voice in their development. I was selected by the AAMA [American Automobile Manufacturers Association] because of my technical background in procedure development within General Motors. (RiS: 1)[29]

These differences in affiliation played out heavily in the working group discussions. One of the Dutch members of the working group, himself employed by a research institute, claimed that bringing down the level of traffic noise was the intention of "people like me who are mainly engaged with assisting in carrying out governmental policies, and who acknowledge that if cars do not become more silent, we will never be able to

28. Personal archives Foort de Roo, File ISO TC 43/SC1/WG 42, unnumbered documents 1993–2001, Member lists 1993 and 1999 of WG 42.

29. The first and former convener of Working Group 42 was Richard Schumacher. The current convener is Douglas More. We interviewed both the former and current convener. Both are affiliated with General Motors.

control the traffic noise problem" (FR: 31). Other members, he argued, were less committed to environmental issues, and had a mind-set that was more focused on trade and economy. In contrast, the ideas promoted by members like the Dutch were occasionally questioned by the "industrial" experts: "I mean, it is one thing to [say]: can you measure something in an—let's say—academic way? It is another thing to make sure that such a measurement can be done repeatedly, that the results can be reproduced, and that they can work for all possible technologies" (DM: 30). The current convener of WG 42, an employee of General Motors just like the first convener, is highly aware of the tensions that may result from members' affiliations:

> How people act personally is always up to them. But in fact you are nominated by your national standards body, and you are representative for them. People sitting around a table are *not* representatives of an organization that may employ them. They are not supposed to be. That doesn't mean they don't act that way.... There is, let's say, how things happen in practice, and how we might wish [them] to happen. (DM: 8)

A Norwegian member had no difficulty in understanding why the Dutch members acted as they did. The Netherlands, he said, is a densely populated area with an extensive network of highways and streets. They were "really looking for good source solutions" geared to tires, road surfaces, and engines (TB: 17). The current working group convener expressed the same opinion: "I mean... you can understand, in a certain context... the priorities that different national organizations may bring in. The Netherlands... by virtue of geography and demographics has more of a concern about noise than almost anyone else. Russia couldn't care less" (DM: 30).

However, despite such understanding, members of the working group deeply disagreed during the meetings. The nonindustry members fostered a system of testing that would measure each single source of noise separately—notably distinguishing tire noise and power-train noise—so that these tests would provide detailed data that could feed into noise control. The industrial members, however, strongly opposed this and promoted a way of measuring that would take all sources of noise together. As two German members noted in one of the working group documents, the automotive industry had always argued that "the vehicle must be tested as a whole." Measurements should not be split up for single sources of sound since the test should provide "a balance of noise sources" as they occurred "in urban traffic."[30] Moreover, the industrial members considered distinct tire and power-train measurements not only costly, but also hard to accomplish:

> Let's just take... that "simple" thing of measuring the engine noise.... This engine tends to be stuck in the middle of something in the front of the car—it has got a lot of stuff around it.... So where do I put my microphones? How do I construct a measure of the noise emission of this engine that's representative when I'm in this very complicated geometric space? Keeping in mind that a lot of things of what humans call "engine noise" might not be the engine at all. (DM: 33)

The industrial members thus phrased their arguments in terms of what was representative regarding urban driving conditions and what was reproducible in tests. The Dutch members, however, were convinced that something else was behind the industrial stance. An integral measurement focusing on the entire vehicle would give the manufacturers much room for maneuver and would minimize the interference of testing procedures in design requirements. As one of the Dutch members put it provokingly: "Total measurement equals total freedom"—that is: freedom for the car manufacturers (BK: 6). Another Dutch expert recalled that members employed by the industry had been particularly keen on protecting this latitude (FR: 36–37; see also US: 4). Porsche, for example, as we know from an interviewee outside ISO but with in-depth knowledge of the car-manufacturing industry, had worked hard to keep its engine sound stable even when it shifted from its boxer air-cooled engine to the six-cylinder, water-cooled system (KG1: 9). Measuring car noise in an integral manner would enable Porsche to continue to do so: it would allow the company to trade one type of noise for another, and keep the sound of the engine as it was.

That the need for leeway in sound design was indeed important to Porsche can also be deduced from a paper presented by Porsche employee Britta Stankewitz at the German Annual Meeting for Acoustics (DAGA) in 2003. Its straightforward title was "Exterior Noise Regulation as Sound Design Constraint." She explained that consumers interested in buying sports cars like Porsche's wanted "attractive," "communicative," and "striking" sounds. Yet maximum car noise limits laid down in international guidelines and national laws hampered both consumers and manufacturers as they tried to reach their sonic goals. More rigorous limits had already "drastically reduced

30. Personal archives Foort de Roo, File ISO TC 43/SC1/WG 42, unnumbered documents 1993–2001, comments on "What the NL delegation agrees," by two German WG 42 members, June 13, 2001: n.p.

the leeway for sound design," and the increasingly high percentage of tire noise in the composition of car noise had made things worse (Stankewitz 2003: 230). Stricter standards of measurement, we may add from what we know about Working Group 42, would further constrain Porsche design.

In the end, those speaking on behalf of the automotive industries won their case: the Dutch gave in, and ultimately agreed to a type of test that allowed integral measurement.[31] This may have been due to the numerical dominance of automotive members in the working group. Another likely explanation, however, is that these automotive members consistently found ways to cast their arguments in the technical language of representativeness and reproducibility. As historians of technology Johan Schot and Vincent Lagendijk have claimed, such semantic strategies are important keys to understanding how engineers in international networks manage to avoid openly political conflicts that reflect national interests. By consistently using "technical," "scientific," and "neutral" language and procedures, engineers have often successfully kept things going (Schot and Lagendijk 2008: 198–99). Indeed, as the current convener claims, it is his task to keep public policy choices out of the ISO tests: "that's the job of the convener: to keep that crap out" (DM: 39). Members agreed that they were engineers, not politicians (PE: 25; TB: 5). As we have seen, several of them acknowledged that politics *did* play its part. Even if several of them had acknowledged that politics might play its part, engineering language helped to keep the working group moving.

Porsche was not the only company that feared the effects of standardization on sensorial design. Rolls-Royce, for instance, claimed that sales figures for their make had fallen dramatically when consumers no longer smelled the familiar scent of the brand's leather seats. Safety regulations had forced Rolls-Royce to make the leather less flammable, yet the leather's chemical treatment had also given it a different smell. After adding a synthetic leather smell, sales had gone up again, as one journalist reporting on the issue noted. (Elich 2011: 5). Standardization was thus believed to be a threat to sensorial design even though Rolls-Royce was able to solve the issue at the level of design itself, while Porsche did so by securing its preferred measurement procedure. Yet it wasn't always as easy as in the Rolls-Royce case to find out which sensorial experiences consumers wanted. The search for the holy grail of acoustics was in full swing. In the late 1990s, the OBELICS project was a major site for this search. What was going on there?

31. Personal archives Foort de Roo, File ISO TC 43/SC1/WG 42, unnumbered documents June 2001, email paper "driving conditions" for WG 42, from a Dutch to a Swedish WG 42 member, November 7, 2001.

The OBELICS research program, launched in 1997, lasted for three years. It aimed at understanding both the subjective and the objective evaluation of automotive sound, and to define "target sounds for different driving situations" (OBELICS 1999: 3, 6). Or, in the words of one of its participants: "What is a sportive car? What should a luxurious limousine sound like?" (WK: 4). The institutes and companies collaborating on the project represented the automotive industry (Renault and Fiat), academic acoustics institutes, such as Oldenburg University, and consultancy companies that specialized in acoustic measurement and analysis, like HEAD acoustics, or in testing and simulation. It turned out to be far from easy, however, to link the results of sound evaluation tests to specifications for preferred sound and actual target sound design.

What made it so hard to accomplish? First, listening to sounds in the testing situations had to be made sufficiently similar to listening to sounds in a real-life situation. As STS scholars have shown, testing depends on *projection*, as it "is assumed that the state of affairs pertaining to the test case is *similar* in crucial respects to the state of affairs pertaining to the actual operation of the technology." Such projections from the small to the large and from the present to the future are often disputed. As Trevor Pinch has claimed, judgments of the projections' similarity and difference "always rest within a broader framework of commitments and assumptions about how a technology will operate," and are "constructed within a body of conventions or within a form of life" (Pinch 1993: 28–29). So which tools would, in the opinion of the OBELICS people, help make test listening and real-life listening resemble each other as closely as possible?

Years before the OBELICS project started, HEAD acoustics, its coordinating company, had already decided that just sitting in a seat and listening to car sounds through loudspeakers or with binaural headphones[32] was not realistic enough. First, HEAD developed a SoundCar that mimicked a car in which the test subjects would not only hear car sounds—via headphones—but would also feel their seat and wheel *vibrating* at the very same time. The disadvantage of the original SoundCar, however, was that when sitting in the test car and opening the throttle, one "would indeed hear the right sounds, but... not feel the acceleration, yet stupidly stand still in the hall" (KG1: 4). As one interviewee added, "It was even a bit comical: we drove about 120 [kmh] and looked out the window, and there were

32. Binaural headphones replay recordings made with the help of two microphones, one for each ear, thus mimicking human hearing.

our colleagues standing next to us" (RH: 3). This is why HEAD acoustics' second intervention was to position the SoundCar on a moving belt and also to transform it into a tool in which the test listeners could actually interact with the car. For this simulation, HEAD acoustics developed its HEAD 3-dimensional Simulator (H3S) software. Instead of being passive recipients of car sounds, the test subjects were now expected to interact with the setting. The headphones or speakers would produce new sounds only when the test driver acted, such as by shifting gears or signaling a turn (HEADlines 2003: 5). The third and most recent step has been the construction of a *drivable* car sound simulator, the Sound Simulation Vehicle.[33] While driving a real car on the road, the driver can hear different sounds channeled via headphones.

Automotive engineers actually working with the Sound Simulation Vehicle, however, such as Ford engineers, stumbled upon new "realisms" when doing the tests. In one such test the sound of a glamorous 500 PS Aston Martin was added to a relatively ordinary make, a Ford Focus. Test drivers opening the throttle of the Ford Focus would hear the sonic feedback of an Aston Martin: "And acoustically you get such feedback that you think, 'Wow, now we're really driving,' but then not much really happens because the car altogether lacks performance." This could be quite dangerous in fact when trying to overtake another car (RH: 7). In this way, the testing situation itself created shifting conceptions of what listening while driving was really about.

If it was a challenge to create testing sites in which the listening situation resembled the *real* situation of listening to car sounds as closely as possible, it was quite another challenge for both automakers and testing consultancy companies to acquire knowledge about what test subjects thought and what they felt about the sounds to which they were exposed. As one of the OBELICS researchers put it graphically, they had to find tools to "squeeze out" the test subject's inner experience of sound as if this subject were an "orange" (NC: 4). For a long time it was not self-evident to researchers that they should use lay listeners when evaluating sound. In acoustics research in general, many early laboratory tests were executed by the acousticians themselves, and it was only in the course of time that the researchers started working with lay listeners. Moreover, the car industry continued to use expert listeners (and still does at present)—often the acoustical engineers themselves—for the testing of particular sounds, such as those from doors (BL; 9; SP: 9; KG2: 19). Using lay listeners became

33. HEAD acoustics developed the Sound Simulation Vehicle in collaboration with Ford's Acoustics Center in Cologne.

increasingly important, however, as the car manufacturers wanted to know how potential customers evaluated sounds.

But how did the automotive industry consider it possible to measure what these lay listeners experienced? Initially, it used so-called pair comparison tests and scale assessments. In pair comparison tests, test subjects compare two sounds (sound A and sound B) that immediately follow each other, for instance in terms of agreeableness, and subsequently compare other pairs: A and C, B and C, and so on. The advantage of this method is that the test subjects do not need to remember the sounds for a prolonged time and can focus on very small differences. Yet, as a HEAD acoustics employee argued, it is "artificial" since in everyday situations "one does not [constantly] jump from one car into another" (NC: 4). Moreover, because the test subjects are "forced" to make a choice, they often feel insecure about whether they "have done the assignment well" (RH: 17), or even start to "think of something else" during the tests (RS: 8). The conventional alternative for pair comparison tests are scale assessments that allow test subjects to evaluate a particular phenomenon with the help of a scale.[34] Again, test subjects find it rather hard to do scale ratings in the case of sound (SP: 6). As one of us experienced herself at the lab of HEAD acoustics, it is difficult to rate the quality of a particular sound without having the option of comparing it to other sounds. Furthermore, what makes one tick the third, fourth, or fifth box on a scale from 1 to 7?

In psychoacoustics research, the semantic differential, a specific form of scale assessment, was introduced for sound perception in 1954. The original test focused on the meaning of underwater sounds to navy sonar operators. It displayed fifty bipolar attributes, such as "powerful–weak," for the evaluation of complex auditory stimuli. Many of these attributes came from descriptions of sonar sounds by students who were taking journalism and had received specialized training in verbal expression and from "sonar recognition cues" used in teaching methods for sonar experts (Solomon 1954: 17–18). Since the 1970s, these lists have been fine-tuned several times, for instance for vehicle acoustics. The members of OBELICS used a set of thirty-four bipolar attributes, such as "boring–stimulating," "cheap–expensive," "loud–soft," and "tiring–relaxing" (Hempel 2001: 170). (See figure 5.2.)

In practice, however, these words were unknown to many of the lay test subjects or were interpreted in a highly arbitrary manner:

34. The Opel YouTube film mentioned earlier shows how this is done in practice; see http://www.youtube.com/watch?v=MJesIniGVjM (accessed January 12, 2012).

("Das Geräusch klingt ...")

dumpf	klar
deutlich	undeutlich
eintönig	vielfältig
Einzelgeräusche ortbar	Einzelgeräusch e nicht ortbar
ermüdend	aufmunternd
gleichmäßig	ungleichmäßig
grob	fein
hart	weich
hoch(frequent)	tief(frequent)
lebhaft	ruhig
rauh	glatt
tonhaltig	nicht tonhaltig
trüb	klar
trübsinnig	heiter
voll	leer

Figure 5.2
Semantic differential
Source: Hempel 2001: 170.
Courtesy: Herbert Utz Verlag.

> For example, what is *rau* [rough]? Do you know what sound *rau* is? What is a rough sound? *Rau* consists of the letters *R* and *A*, rrrrrrrr, is this rough? Is aaaaaa rough? [I (KV) reply: "Rather R than A."] Psychoacoustically, though, exactly the opposite is the case; A is very rough psychoacoustically because it comes with a high intonation. R comes with a deep intonation. It is not at all as rough, from a psychoacoustical angle, but in practice you are right, for most people think of *Rauhigkeit* as rrrrrrrr. (KG: 1, 11)

Moreover, during the semantic differential tests, the researchers noticed that the test subjects were inclined to tell the test leader much more than was actually asked for. One subject insisted on telling about her vacation experiences while traveling by airplane since she thought the car sounds were "similar" and equally "pleasant." Another subject left the test site in tears "because a specific sound reminded her of an accident" (NC: 4). In order to record these additional stories, the researchers introduced Associated Imagination of Sound Perception (AISP), which was later reworked into Explorative Vehicle Evaluation (EVE).

In the AISP tests, the subjects, while seated in a SoundCar, were asked to comment on sounds freely and mention every association they might think of. Subsequently, they were interviewed about their comments, and all of them "found sentences to express themselves" (NC: 19; SP: 5). Through such tests the researchers claimed to reach "another dimension in the description of noises," one much closer to the test subjects (NC: 5–6; AF: 3; RoS2: 4). Instead of mere evaluations of sound, one now had a way to understand their origins—the *why* behind the *what*. Each person's assessment of a particular sound, they said, was based on *gesammelte Erfahrungen*, a series of earlier experiences, and memories and expectations built upon them. Two persons, for instance, might both consider a car sound loud. Yet, while the first one would love it, claiming to be a "fan" of the car tested, the second considered it "too loud... for I imagine myself having to drive this car for five hours, which I do regularly, Paris–Aachen, for instance" (NC: 6). The explorative vehicle evaluation tests were tailored to elicit such stories in situations in which test subjects would actually drive a car while listening.

However, the usefulness of the AISP and EVE tests for sound design was highly contested within the automotive engineering and acoustics community, and not only because the tests were time-consuming and expensive (WK: 13; RoS2: 3). Scientists and engineers from Oldenburg University, for instance, considered the tests far too vague and doubted their usefulness (NC: 12–13; KG1: 14). Many acoustical engineers had hoped to set up a sort of universal sound quality index, in which a series of numbers would indicate which design specifications would lead to sounds of high quality (RoS1: 4–5). They did not see how the varied outcomes of the AISP tests could bring the sound quality index any closer (KG1: 14–15). An Opel engineer explained that he actually preferred a sound-quality system that would get rid of differences in taste (BL: 2). Many of the OBELICS personnel involved in subjective testing considered such a sound-quality index a bridge too far. In their view, sound quality in a car simply depended too much on context (TH: 3; WK: 10; RoS1: 6; SP: 14). Several of them claimed that one could define preferred sounds only for very specific groups of consumers. This implied that one should also invite narrowly defined target groups for the testing procedures (NC: 8; RS1: 10; SP: 11). HEAD acoustics' president Klaus Genuit therefore felt that AISP and similar tests could be commercially exploited only for quite specific versions of target sounds or for "benchmarking" (KG1: 3).

Although Genuit expressed disappointment over the engineers' inclination to consider everything measurable (KG1: 14), he is an engineer himself—as are many OBELICS members, even those involved in subjective testing. This is why the line between proponents and opponents of AISP was never simply one between engineers and nonengineers. Rather, there was, by and large, a distinction between nonengineers and engineers involved in *consulting* on sound design on the one hand and those involved in *implementing* sound design on the other (WK: 10–11). Given their reliance upon numbers responsible for implementing new designs, the "implementation" engineers considered AISP not easily "management-compatible" (TH: 5).

Some makes, such as the Harley Davidson motorcycle and the Porsche sports car (RoS1: 12, 15; PE: 3), already had exterior target sounds connected to their image. Yet in these cases, several interviewees claimed, engine sounds originated first, while the idea to treasure these as target sounds came only much later. Such "acoustical fingerprints" might work (RH: 14). Still, designing a target sound from scratch was widely considered to be a completely different challenge. What's more, who should have the final word in the actual design? A Ford employee explained that one simply could not leave the conception of target sounds in the hands of the test subjects only:

> When you ask customers how a sound should be, you will get as many different answers as the number of customers you ask. This means that in part you cannot have the customer decide; after all, you need to hold on to the characteristic Ford sound. As a distinctive sound it also has to be embedded in a whole range of different things. The sound should fit in with everything we want to convey through our car philosophy. (RH: 8)

In the ideal world of this Ford sound engineer, future Ford customers would be able to download a series of car sounds (for the turn signals, seat-belt warning, windshield wiper, et cetera) of their choice, just as they would upload a ringtone for a cell phone. Yet, and this was a crucial twist, this option should be available to the customers only after the Ford sound engineers had created a full "sound composition" in which all of the sounds would both be typically Ford and go together extremely well (RH: 11–12; PE: 4). In addition, design needed to be accompanied by marketing techniques. When Ford designed a new fancy turn-signal sound ("pock-pock-pock") for its Ford Focus, many test subjects and automotive journalists initially rejected it owing to nostalgia for the venerable "click-clack" sound originally linked to the relay. Only after the Ford Focus had been advertised as a first-rate modern car and a story had been linked to its futuristic turn-signal sound did the new sound come to be accepted (RH: 9–10).

Even discounting the sonic cultural conventions the automakers had to deal with, it is clear that in the debate over sound quality, acoustical engineers desiring an index and researchers heralding the significance of subjective testing by lay listeners were pitted against each other. One way to understand the tensions between the two is suggested by the work of linguistic anthropologist Tom Porcello (2004). Porcello analyzed a conversation between an experienced, professional producer and a student, a novice, in a recording studio by focusing on their talk about sound during a recording session. While the professional was used to talking in terms of musical instruments and technologies such as microphones to explain the kind of sound he wanted, the student referred to particular bands and songs to express the sounds he preferred. This led to profound misunderstandings between the two despite the producer's efforts to bridge the gap by using metaphors and mimicry.

These misunderstandings seem to be similar to those between the index-loving acoustical engineers and the protagonists of subjective lay testing, with the test subjects in the role of novices and the acoustical engineers in the role of the producers. The discursive clash that Porcello analyzed also reminds us of the misunderstandings that originally kept apart aircraft engineers and pilots in a flying qualities case described by Walter Vincenti. In order to explain this example we should first bear in mind that despite the deep disagreements about testing just described, most testing is "routinely accepted as fact," as Donald MacKenzie has phrased it (1989: 415). His colleague Edward Constant has shown that "traditions of testability" are at the heart of such routine practices of communities of technological practitioners, and that accepted methods of testing are closely interconnected with the dominant design characteristics of the technologies they work on. This implies that introducing a new technology in a particular industry usually also requires "the creation of new or much refined testing techniques" (Constant 1980: 22; see also Constant 1983).

How creating a new tradition of testability works in practice has been illustrated by Vincenti in his book *What Engineers Know and How They Know It* (1990). In one chapter, he examines the establishment of design requirements surrounding the flying qualities of American aircraft in the interwar years. The idea behind the specification of flying qualities was to produce aircraft "possessing adequate stability, responsiveness to the controls, lacking in eccentricities or sudden changes of behavior, and generally satisfactory to the pilot" (consulting engineer Edward Warner, quoted in Vincenti 1990: 81). The initial problem, however, was that test pilots expressed their opinions of flying qualities in highly subjective and qualitative language. In their terms, the controls would respond "very readily," for instance, while the pressure exerted on the controls was "normal," and the longitudinal, lateral, and directional stability were considered "good" (Vincenti 1990: 64). Yet this bore no relation to the formal engineering criteria for describing particular types of aircraft stability, or to what the engineers could measure in flight. Only after prolonged close cooperation between engineers, instrument makers, and pilots at a particular laboratory did the aircraft industry manage to develop a set of precise instruments and piloting techniques to make measurements in flight, as well as achieve increased refinement of pilot opinion, so that the pilot "*knew* what he liked" and "could *say* so" (Vincenti 1990: 103, our emphasis). In this process, the use of "standardized terminology and definitions for research pilots and engineers and standard rating scales for quantification of pilot opinion" was paramount (Vincenti 1990: 98). This shared vocabulary would not have been possible, however, without the iterative testing in which small groups of experienced pilots, instrument makers, and engineers—expert testers—collaborated for long periods of time.

In the car sound evaluations, however, there was little sustained collaborative effort between the engineers and the lay listeners. We explain this by referring to the marked differences in status between test pilots and test drivers. While the first are members of a profession with exclusive jurisdiction concerning the flying of aircraft, everyday drivers cannot claim such a special position. The officially acknowledged exclusiveness of the pilots' skills, and certainly *test* pilots' skills, fueled the interest of the engineers in a sustained collaboration with these pilots (Vincenti 1990: 70–71).

We have already mentioned that the acoustic consultants at OBELICS considered it important to work with lay listeners. Preferably, these lay listeners would be highly specific groups of consumers tailored to the type of car tested. As one engineer said, it was quite absurd to have the sound of cars worth a hundred thousand euros assessed by students—not exactly the usual candidates for driving such cars (BL: 3). At the same time, however, many of the interviewees stressed that expert listeners noticed far more than lay listeners (RoS2: 5; WK: 14; BL: 5; KG2: 13). In contrast to lay listeners, experts were able to determine in the laboratory which sounds would sound *natural* outside the lab's walls (BL: 6–7). Yet outside the lab, the interviewees preferred lay listeners. Why had this acquired such crucial significance?

MARKETING SOUND IN THE EXPERIENCE SOCIETY

As we have seen, increasingly stringent limits for the maximum sound level emission of cars affected their sound design. Both the car industry and cultural sociologists, however, tend to explain the increased focus on sound design by referring to the rise of issues related to the "experience society." German sociologist Gerhard Schulze (1992) argues that sensory experience plays a large role in the selling of consumer goods today. This emphasis has emerged because many products have been perfected to such a degree that differences in technical specifications between brands have decreased over time. Since the 1980s most products now simply *work*. Moreover, most consumers today have an enormous array of consumer goods to pick from on account of the booming postwar economy. The many options available, however, also bring about uncertainty. To compensate for the absence of difference in technical quality and to make choosing easier for consumers, products are increasingly sold by cashing in on the emotional significance and inner experience they evoke in buyers.

In the marketing of new products, therefore, sensory experience has become crucial. What appliances feel, smell, or sound like and how that matches the buyer's identity have thus become as relevant as how they look (Marks 2002: 114; Wenzel 2004). All this has led to a strong "aestheticization" of everyday life, described by Schulze as the emergence of the *Erlebnisgesellschaft*, or experience society, and by sociologist David Howes as the "sensual logic of late capitalism" (Howes 2005: 293). In such a society, experience value trumps functional value both in selling strategies and in the motivation of consumers (Schulze 2005 [1992]: 59). Schulze is critical of experience society's focus on immediate and subjective inner experience (*Erlebnis*) since that may come at the expense of a more long-term and well-grounded experience transferred from one person to another (*Erfahrung*). His work, with its focus on abundance, has been criticized for neglecting new forms of scarcity as well as signs of being oriented to the world beyond the inner self in particular milieus, such as the elderly (Müller 1993).

Yet many of our interviewees within or related to the automotive industry presented an analysis of today's society similar to that of Schulze (KG1: 9; RH: 20–21; BL: 10–11; AF: 4; RoS2: 8). Since there are many wealthy people these days, one interviewee claimed, their concern with a high standard of living has come into play. In addition, since they no longer have to worry about whether their cars actually run, they might as well complain about their sound (NC: 11). What's more, the automotive industry considers sound more easily observable by consumers than other aspects of automotive quality, such as safety (Zeitler and Zeller 2010 [2006]: 1). Dependent on customers' specific preferences, cars should have "decent," "luxurious," "dynamic," or "sporty" sounds (Bernhard 2002). Renault, for one, made its Clio 3 "silent" and gave its Megan Sport a "dynamical sound" (EL and VM: 2).

Although there may not be sufficient evidence to warrant an analysis of contemporary culture in terms of the experience society, what is relevant to our understanding of the tensions between subjective car sound evaluation and the search for objective car sound specifications is that the world of marketing *believes* in such a society and in sound as a "marketable vehicle attribute" (Repik 2003: 5). Designers are now obliged to "ascertain what emotional values they want the consumer to attach to the product. They then develop forms which instigate the associations to, hopefully, inculcate those feelings" (Guy Julier, quoted in Lury 2004: 87). Or, as brand experts Hajo Riesenbeck and Jesko Perry explain, fully in line with Schulze's analysis, "The price of a new car is so high that buyers will have the basics covered anyway. . . . This makes it imperative for automotive advertising to appeal to the heart and soul rather than reason" (Riesenbeck and Perry 2009: 29). This has grown even more important after an increasing number of different car makes began to make use of the exact same set of car parts (Gottfredson et al. 2001).

This shift in marketing strategies needs qualification to prevent misunderstanding. We do not claim that until recently marketing people had no eye for things such as the *image* of a product or the *lifestyle* to be associated with the product. On the contrary, Wolfgang König has convincingly traced attempts to create a positive product image by means of symbolic messages in advertising in the years prior to World War I (König 2000: 405). Nor do we claim that companies did not have their products tested by consumers: In Western Europe, firms started establishing testing labs and design departments in the 1950s (Oldenziel, de la Bruhèze, and de Wit 2005: 118). Yet, while manufacturers had long been interested in what *the consumer did with the product*, they have increasingly become obsessed by what *the product does to the consumer*. Manufacturers now want to have tested which emotions their products stimulate in test subjects, the future consumers.

It is in this context that an extensive research industry has evolved that concentrates exclusively on developing testing methods for tracing these feelings. One component of this new research industry is HEAD acoustics, providing the automotive industry (and increasingly the domestic appliances and personal computer industry) with testing methods and actual testing of sound. In turn, companies such as HEAD acoustics collaborate with universities, whose staff members equally address the changes associated with today's experience society (Schulte-Fortkamp, Genuit, and Fiebig 2007: 12). The notion that sound "is well known to enhance or detract from our pleasure in possessing or using a product" (Boulandet et al. 2008: 1) has thus been reinforced by an emerging and growing network of manufacturers, designers, testing companies, marketers, and academics who reciprocally *spread the word* of sensorial branding and design.

In light of the trends in marketing just mentioned, it is understandable that HEAD acoustics endeavored to develop qualitative tests that would enable the full expression of the rich variety of test subjects' emotions with regard to car sounds. Because these tests had to reflect the evaluation of untrained listeners, that is, the everyday consumer of cars, HEAD acoustics increasingly selected lay listeners. It is ironic that while the automotive industry considered these consumers highly sensitive to sound, it took new ways of testing to have these consumers properly express their assessments of particular sounds. The experts' idiom of sounds did not speak to these lay listeners. The new tests in the OBELICS project did not straightforwardly lead to the specification and design of target sounds, however. Nor did it generate a sound quality index. Yet understanding how consumers perceived the sounds of cars remained an important issue for psychoacoustic research.

SENSORY INSTRUCTIONS

In the 1980s and 1990s the automotive industry displayed a widespread belief in the rise of the new *experience society*. This resulted in new marketing strategies that highlighted the significance of car sounds. The driver's emotional experience of quality car sound, so it was believed, would help sell a car. Moreover, specific target sounds would enable new connections between a car make and particular groups of consumers.

This belief led to an emerging *tradition of testability* in the European automotive industry. The crux of this testability, in which the German company HEAD acoustics played the lead role, is subjective testing, which enables free, associative verbalization of car sound perception. Moreover, testing car sounds while driving (in a mobile SoundCar) became the heart of *projection*: the idea that the testing was "similar in crucial respects" to actual driving.

The OBELICS project, in which representatives from the automotive industry, acoustics departments at universities, and testing consultancy companies collectively aimed at linking subjective testing to objective target sounds, did not immediately lead to the definition of such target sounds, however. In contrast to a similar research project on the specification of flying qualities for airplanes, and in contrast to previous habits in the automotive industry, HEAD acoustics used *lay* rather than *expert* testers. Although this was fully in line with the new marketing strategies of the automotive industry, it did not readily facilitate fruitful collaboration between automotive engineers and sound designers. Whereas test pilots and aircraft engineers had found a common language for the design specification of aircraft flying qualities through long-term, close, and inter-professional collaboration, the lay testers expressed their sonic preferences in the absence of the designers. From the designers' perspective, the lay testers behaved like the novice in Tom Porcello's study of studio engineering: The words they employed for sound were incomprehensible to the sound designer.

We should not forget, however, that despite the difficulties in directly translating consumers' auditory experiences into sound design, car manufacturers kept working on the acoustics of their products. Their consistent and continuous focus on how their cars sounded, even though they could not always lay their fingers on what consumers exactly wanted to hear, only underlines how significant car sound design had become within the automotive industry. The same holds true for their investments in making sure that *standardization*, aimed at facilitating international trade yet in need of responding to new societal and technological trends, would not thwart their options for tinkering with their brand-related car sound design.

The automotive industry's investments were not restricted to examining drivers' sonic preferences or designing target sounds, they also had to find ways of actually selling the sound to their clients. In 2002, STS scholar Michel Callon and several colleagues assumed that in an "economy of qualities" consumers are "capable of perceiving differences and grading them" (Callon, Méadel, and Rabeharisoa 2002: 212). In practice, however, marketing people help consumers to perceive the nuanced differences between the products they offer through *grade labeling*, a "system of identification that describes products by their quality" (Medina and Duffy 1998: 229). In case grade labeling involves intangible attributes, it will often require the standardization of sensorial experiences such as smell, taste, and sound. For instance, our supermarket may teach us that a label with the color yellow stands for a particular, well-defined taste of wine.

Even such grade labeling may not be sufficient to convince the consumer. As sociologist Eva Illouz has explained in her work on *emotional capitalism*, selling the nuanced sensory characteristics of products has required an increasing fine-tuning of marketing strategies. While the eighteenth century already had rich verbal descriptions of colors such as "Turkey red" and "Prussian blue," today's marketing people are seeking novel ways to evoke and capture sensations. "Consumer fantasies are elicited by the ways in which advertising evokes the sensory character of goods," Illouz claims (2009: 403, 405). In her view, advertisements do so both through "perceptual mimesis" and "delayed sensory content," concepts she has taken from literary scholar Elaine Scarry. While fictional narratives in advertising have no actual sensory content, they trigger in the consumer sensory experiences through perceptual mimesis. A reader may hear a smooth-driving car while reading a description of such a car. Yet ads also have "delayed sensory content" in that they provide "*instructions* for the production of actual sensory content as with a musical score" (Illouz 2009: 404, our italics).

We should thus read television commercials like the ones for the Peugeot 508, Toyota Avensis, Volkswagen Passat, and Mercedes-Benz we mentioned earlier as *musical scores*. Although the sounds and images of such commercials have immediate sensory content, they provide delayed sensory content as well. The ways in which these ads notate how men find sonic solace in their cars after and during their busy working days, are instructions for sensory content that drivers are considered to produce as soon as they are close to actually buying the advertised cars. Just like musical scores are notations for evoking musical experience, the ads are scores for how we, as consumers, should "feel" particular cars and experience the sounds they bring with them. Whether or not this always works as intended is something we do not know. Yet from our previous chapters we *do* know that drivers have used the sonic characteristics of their cars, both their cars' acoustic isolation and auditory features, for acoustic cocooning, for feeling sound and safe. And by now we also know that the automotive industry's search for the sounds that best capture the identity of their makes, as well as the strategies they developed to get there, can only be understood if we acknowledge that the 1990s were the age of belief in the experience society and in emotional capitalism, of increasing standardization, and of a newly emerging tradition of testability.

CHAPTER 6
Auditory Privacy and Sonic Relief in the Car

UNDISTURBED

"Undisturbed" is the title of a short item published in the Dutch *Financial Newspaper* of August 21, 2010 (Anonymous 2010a). It discusses a new type of car radio recently launched by Kenwood: the KDC-BT60U. Its "Highway Sound" function is designed to prevent the nuances of radio music from being drowned out by highway noise, while its "Supreme Technology" is meant to ensure that the sound quality of music from MP3 files equals that of music on compact discs.[1] As blogger, technologist, and entertainment sector businessman Dwight Marcus claimed in 2009, the game of recorded music as "soundtrack for the living" is changing. "For most of the listening day, our audience is not in an acoustically-treated control room, a pin-drop-quiet concert hall, an intimate acoustic music club, or bathed in a bone-crushing dance-hall system." The new ideal is to "keep music sounding natural but subtly enhanced under pristine listening conditions and subjectively excellent in a car on a noisy freeway, or while playing from another room during animated dinner conversation."[2]

In the early 1980s, high-end audio companies like Bose had already started to equalize the sounds of the car's audio systems and to use increasingly powerful amplifiers and precisely placed speakers. At the turn of the century, digital signal processing boosted fine-tuned equalizing, and Audio Pilot "automatically" adjusted "sound levels to counteract ambient noise."

1. http://www.kenwood.nl. For more details, see http://www.kenwood.nl/products/car/receivers/bt_receivers/KDC-BT60U/details/ (accessed January 30, 2012).
2. AurumAurus, Car as Concert Hall, posted March 24, 2009 at http://dwightmarcus.wordpress/com/2009/03/24/ (accessed May 17, 2011).

Subsequently, surround sound systems like the Dolby Digital 5.1 or Logic 7 widened the distribution of speakers across the car and digital satellite radio extended the choice of radio program channels (den Hoed 2005: 3).[3]

The investments in quality speakers and the remarks on "pristine" listening conditions in the car warn us that we should not take advertising of acoustic cocooning so seriously that we assume ideal listening conditions available in each and every car. Manufacturers of midrange and luxury car makes were most active in silencing their cars' interiors. But the rise of both noise-sensitive and highly receptive car audio systems does signal a new phase in audio development in which acoustic cocooning *within* the car paralleled seeking connection with the world *outside* the car. The use of mobile office technology and navigation technologies (Pels 2010), the introduction of features for synchronizing music listening at home and in the car (Schenk 2011), and the rise of in-car social media enabling drivers to connect with fellow drivers by sharing, for instance, music (Östergren and Juhlin 2006) equally indicate a combined focus on acoustic cocooning and (often sonic) connecting. Whatever drivers' sonic activities in the car—listening to their radio, enjoying the relative tranquility of the interior, feeling comfortable amid its unique sounds, or speaking on a hands-free mobile phone, they are still largely controlling who and what is sonically able to enter the car. It is like being at home, but with more physical privacy than home usually offers. Over time, one might say, the car has not only developed into a more or less convincing version of a "concert hall on wheels" (den Hoed 2005: 1) or a mobile living room, but also into the twenty-first-century version of the nineteenth-century "cabinet": the small, private, luxuriously furnished and embellished room in which the elite enjoyed after-dinner time and discussed the business affairs that required some discretion.

This final chapter will bring together various historical understandings of how drivers found auditory privacy and sonic relief in the car. We will then return to an issue already announced in introduction, but now pressing for an answer: if the contemporary appreciation of car sound is hard to capture for researchers, how, then, are we able to gain access to the connections between driving, listening, and feeling at home *in the past*? Even if we succeed in clarifying this issue convincingly, two questions remain to be answered. What about the boom car, a vehicle equipped with an extremely powerful audio system—doesn't that contradict our conclusions about

3. Ivan Berger, "Driving; Car as Concert Hall: Audio on the Road," *New York Times*, April 12, 2002, at http://www.nytimes.com/2002/04/12/travel/driving-car-as-concert-hall-audio-on-the-road.html (accessed May 17, 2011).

the trend to acoustic cocooning? And finally, what are the implications of acoustic cocooning for our notions of public and private space, and for the issue of responsible behavior on the road?

A HISTORY OF LISTENING BEHIND THE WHEEL

If anything has become clear from the previous chapters, it is that we, as drivers, have found auditory privacy and sonic relief in the car without actively seeking it, at least initially. This is not to say that manufacturers just invented the needs of drivers; rather that automotive engineers often acted as mediators between what they knew about consumers' wishes and what they considered efficient manufacturing. What we would like to underscore is that manufacturers and engineers often considered noise control and sound design solutions to problems that had little to do with our contemporary notion of seeking moments of tranquility in our often fast-moving and privacy-depleted lives. Listening behind the wheel came in many flavors.

Automotive sound has in fact been part and parcel of design and marketing ever since the 1920s, even though it took on a new significance and twist in the 1990s. Engineering sound and vibration became particularly relevant in the interwar years. In the car's transformation from motorized open coach to sealed box on wheels, engineers, marketers, and drivers started to listen to the sounds of cars in new ways. This process toward the closed sedan expressed a shift from the car as a toy for rich, upper-class, and adventurous sportsmen to the car as a vehicle for middle-class businessmen and for families who used their automobile to go on country outings. It also marked a development in which the production of car bodies, originally the craft of independent coachbuilders, increasingly became embedded in car manufacturing itself, notably after the introduction of the all-steel body in the mid-1920s—succeeding fabric bodies and semirigid wood and metal composite bodies—and the design that combined chassis and body into a single unit in the early 1930s.

In the early twentieth century mechanical engineers already began to consider machinery noise a sign of wear and tear that indicated energy-absorbing friction and loss of engine power. Hence, in the automotive industry and elsewhere, the ability to silence mechanical noise came to stand for engineering quality as such. Silencing the engine—by balancing its rotating components—and the gearbox were priorities number one and two. In addition, manufacturers used rubber, leather, and cotton-reinforced Bakelite to prevent transmission of vibrations from one component of the car to another, since closed-body cars were far from silent. While closed bodies made sounds audible that had been previously masked by the sounds of wind, traffic, and air resistance in open cars, the composite bodies had specific squeaks and rattles, and the all-steel bodies initially worked like boom boxes, amplifying the vibrations from chassis and drivetrain. Silencing also served a novel goal, however: enhancing the *auditory impression* of driving a *reliable* car, as we know from sources in France. Tailoring the car for middle-class families, including women, meant automating processes that had previously required manual work and making sure that drivers trusted their cars. Quieting the car helped to avoid noises that could mislead and worry inexperienced motorists or lure them into incorrect diagnoses of engine problems.

Engineers sought not only *mechanical* silence, however, but also *convenient* silence. By this we mean that they hoped to create cars that were both less prone to mechanical fatigue *and* would cause less nervous fatigue in drivers. It was in the interwar years that physicians and psychologists began to consider noise as potentially impeding the productivity of urban white-collar workers. In the same vein, automotive noise was seen as an exhausting force that might easily sap the energy and vigilance of drivers. As to the car body, this meant that manufacturers wanted mechanical silence not just in the sense of a body keeping silent but a car body that provided convenient silence: a silence within the interior space of the car that insulated and protected occupants from outside noise. Shielding technologies such as layers of felt in addition to the floor panel between the chassis and the body provided the answers here. Such convenient silence could compensate for a lack of mechanical silence of, for instance, the chassis. Test drivers indeed gave testimony of reduced fatigue when driving such noise-insulated cars.

Finally, car manufacturers promoted what we have called *aristocratic* silence: silence as a sign of the car's and the driver's societal standing. While for a long time speed had been seen as the aristocracy of movement, silence became the new aristocracy of speed. Advertising thus tapped into several age-old symbolic varieties of the "aristocracy of silence." While elites often had control over who was allowed to make sound—forcing those lower in hierarchy to remain silent—and were wealthy enough to keep themselves aloof from everyday noise, "noblesse oblige" also increasingly implied the capacity to control oneself and to keep silent if circumstances required this. All these associations between aristocracy and silence can be found in the car ads of the 1920s, 1930s, and 1940s. Manufacturers advertising the silence of their cars elicited connotations of aristocracy through references to luxury items such as rustling silk—wealthy silence—or royal coaches: an unimpeachable space. They also invoked elegant and silently moving animals such as swans and panthers. And they referred to delicate ears—a sign of refinement—and the silence of shadows: not only the literal silence of the shadow but probably also the silence of servants.

This pursuit of mechanical, convenient, and aristocratic silence was perhaps more reflective of an ideal than a reality. But the silencing effort also helped sustain something completely different: the highly valued cinematic experience of the driver. Our study of European literary sources showed that in the first decades of the twentieth century literary motorists described driving as being elevated above everyday life, experiencing the world as being more distant and less intrusive than normally, similar to the experience of watching a movie. The closed car body impeded this experience, however. Inadequate ventilation misted up the windows, and surrounding metal blocked the occupants' views. In addition, balloon tires enabled increased speed, which in turn brought more noise and vibration. All this, however, reduced the experience of a smooth, cinematic drive. It was exactly for this reason, as we have come to know by following American engineers, that the automotive industry started to study the multisensorial perception of noise, vibration, and harshness by drivers, and invested in improvements in suspension systems and tires. This effort contributed to preserving the tourist gaze associated with early automobility. The idea was to take any road in such a way that the only mark of motion would be the passing landscape. Issues of sight thus help to explain the early rise of car sound design. Later, issues of sight sustained (though not caused) the use of in-car audio entertainment when the driver's view became increasingly blocked by road noise barriers in the 1980s and after.

As we have seen, one of the reasons why engineers fought noise in the 1920s was to create an auditory impression of reliability. In fact, monitory and diagnostic listening to engines was considered drivers' everyday duty in the first decades of the twentieth century. Such modes of listening constituted drivers' "acoustemology" or "techoustemology," to repeat the terms coined by Steven Feld and Thomas Porcello (mentioned in chapter 1). Diagnostic listening, however, gradually came to be seen as something that should be the exclusive expertise of mechanics. This was especially true for Germany, where car repair evolved into a protected, guild-like profession in the 1940s, but an analysis of European manuals for motorists published between the 1930s and the 1970s provides evidence of such a development as well. While the early manuals gave elaborate descriptions of sounds and how they could inform the driver of troubles, later manuals included less and less of such auditory information.

Intriguingly, this process of *delistening* to the engine—as pertaining to drivers—was paralleled by a process of increasing *car radio listening*, a process we have examined for the years between the 1920s through the 1970s. The meanings attributed to car radio shifted over time, however, and in telling ways. Early radio was "sold" as a companion to the lonely driver, while one engineer even suggested radio was a way to mask the noise audible in a closed sedan. From the 1960s onward, however, radio increasingly came to be defined as a sonic assistant helping drivers to cope with their *lack* of solitude on the road: to musically control their temper in traffic jams and during encounters with irresponsible fellow drivers. The listening driver was thus *encapsulated* in new ways: not only detached from the sound of the engine, but also guided by the mood-improving sound of the car's audio equipment. In addition, the 1970s saw the rise of traffic information on the radio, which enabled drivers to find alternative routes when they were stuck in traffic. It was no coincidence that car radio was advertised as a "magic carpet," as it could help the driver to keep moving with the help of traffic information or experience movement in the rhythm of music while trying to cope with jammed roads.

Drivers would have still more novelties to deal with: the noise barriers that started to encapsulate both cars and roads in the 1970s, 1980s, and after. They were designed to protect citizens from noise, yet they also blocked drivers' view. And whereas highways promised freedom of movement, noise barriers were reported to have an imprisoning effect on drivers. Our argument has been that we can best grasp the contested character of the noise barrier (and its effects on acoustic cocooning) by understanding it in two different ways. Socially, we interpret it as a specific type of NIMBYism—with the driver treating the landscape as his "mobile home's private garden." Culturally, we see it as an expression of deeply rooted fears and experiences that bear a remarkable relation with ancient tropes of the "underground," as described by Rosalind Williams in her study of eighteenth- and nineteenth-century literature on caves, mines, and tunnels. Drivers have described these experiences as claustrophobia, a beneath-the-earth experience, a gutter feeling (of *being* driven by a stream stronger than themselves), and an involuntary separation from the natural landscape. Drivers see "barried" roads as buried roads, a classic underground theme, while the experience of being caught in a deep drain with endless walls reiterates a more sublime experience: that of the "artificial infinite," of big structures that force drivers into submission.

Noise barrier designs that responded not only to environmental concerns but also to the trope of the private garden were met with most enthusiasm by drivers: they liked green and transparent noise barriers.

In general, however, the many miles of barriers continued to be seen as a sensory deprivation forced upon drivers, a deprivation for which they could seek compensation only *within* their car. This made the car into what Williams has identified as a fantasy of a fully enclosed, silent, and magic space: a more recent underground trope that did not find its materiality in the design of roadside barriers but in the design of cars. It was the car itself that became a refuge, with artificial temperature, light, and sound, a magic carpet compensating for the sensory loss of the noise barrier road. Car radio and stereo equipment once again took on new functions and meanings: in radio broadcasting, programs tailored to drivers became important. Moreover, drivers increasingly used their audio systems to play self-chosen music and audiobooks—new forms of acoustic cocooning that assisted the driver in regaining control, or a sense of control, on a corridor that largely controlled the driver. Through education, entertainment, and imagination, car audio could enliven the most boring commute.

Finally, we have explored the rise of a new tradition of sound design and testing in the European automotive industry in the 1990s and beyond. The 1970s national framework laws on noise abatement not only brought noise barriers, but also forced the automotive industry to take increasingly stringent maximum levels of noise emission into account. Moreover, the subsequent European harmonization of such legislation required standardized ways of measuring car sound. Such standards, however, threatened to reduce the automotive industry's freedom to decide which components it wanted to silence, precisely when sound design acquired the new significance we suggested earlier. This 1990s sound design focused on creating brand-specific sound and target sounds, that is, those designed for specific groups of consumers. We showed that the automotive industry went to great lengths to secure design freedom in the context of the International Standardization Organization. We also discussed manufacturers' investments in research and development concerning sound and how they struggled to testing sound in a way that could really open up how drivers evaluated it. We argued that the rise of the experience society, or actually the automotive industry's *belief* in the experience society, helped us understand why the industry has become so keen on personalizing the car by offering every type of consumer a different *choice* of sound—thus directing the driver once again toward a new form of acoustic cocooning.

The rise of acoustic cocooning has thus been a history in which shielding the driver from noise, filling the car with audio and designed sounds, and connecting the driver with self-chosen informative sounds have been interrelated. Both the driver and the car have become encapsulated, acoustically and visually, literally—through sound insulation—and, in the case of the driver, also metaphorically through the emotional protection by car audio. This last issue also articulates how the driver has not only been shielded from noise but has also been offered sonic alternatives. The car itself still speaks to the driver, but largely through the designed sounds of warning signals and the sounds of solidity, much less as a machine that directly expresses its state of well-being through noise. We have to be careful, but we have reasons to believe that contemporary motorists listen to their cars differently than their predecessors did: less analytically, and more in search of sounds that match the driver's emotional needs. Psychologists thinking about sound design increasingly approach car sound in a similar way. They, for instance, conceptualize sounds in the car, such as the computer voices of navigation and additional driver support systems, as sounds that should adapt to the emotional state of the motorist in order to heighten the driver's attentiveness and increase automotive safety (Nass et al. 2005).

The car is now an acoustic cocoon, a magic carpet and a cabinet in which we can find sonic relief and in which we cherish our auditory privacy. The road to being sound and safe, our stories showed, has expressed the world outside the car as much as the individual motives of drivers. So far our analysis, our history of listening behind the wheel. But how did we acquire access to drivers' sensory experiences of the car in the past?

MOVING, SEEING, LISTENING—IN THE CAR AND IN THE PAST

Fierce debates among scholars in the humanities are not as frequent as one might expect or even hope for, but the world of sensory studies recently had one. Sarah Pink, David Howes, and Tim Ingold, all anthropologists, openly clashed about the best way to study the senses in culture. Basically, Ingold blamed Howes for not really targeting sensory *experience*. In his view, anthropologists of the senses only focus on the sensory hierarchies and modalities within particular cultures and how these inform people's *metaphors* of sensory experience. They examine how people cognitively assemble sensations in terms of "received cultural categories" (Ingold 2011a: 314) rather than studying how people make sense of the world while being *engaged* with their environment through moving, watching, listening, touching, and smelling—which would be Ingold's preference. Howes, in turn, has reproached Ingold for promoting a phenomenological and nonsocial version of sensory studies that neglects indigenous experiences and understandings (Howes 2011; see also Howes 2010a, 2010b).

Sarah Pink lines up with Ingold in pleading for a sensory anthropology that is informed by theories of perception stemming from areas as diverse as neurology, geography, phenomenology, and practice theories, even though "the study of culturally constructed sensory categories" might be included (Pink 2010a, 2010b: 337). All agree that sensory studies should take intersensorial relationships into account, but disagree on how this should be done.

It all started when Ingold published "Stop, Look and Listen! Vision, Hearing and Human Movement" in his book *The Perception of the Environment: Essays on Livelihood, Dwelling and Skill* in 2000. In this essay, he develops a two-step critique of the anthropology of the senses. Western thinking, he claims, has been haunted by the idea that seeing and hearing are radically different modalities, in which the visual is associated with a static position that distances itself from the world, objectifies, and defines "the self in *opposition* to others," while the auditory is considered to be dynamic, intimate, subjective, and defining "the self in *relation* to others" (Ingold 2000: 247, his italics).[4] Yet instead of breaking down the divide between seeing and hearing, anthropologists of the senses have done much to reify it. They have focused on comparing cultures in terms of "the relative weighting of the senses through which people perceive the world around them" (Ingold 2000: 250). They not only showed that some cultures are more oriented to the aural than is the West—at times implying a moral superiority of the people within those cultures—but also suggested that seeing has less priority in these societies, despite empirical evidence of the contrary.[5]

That anthropologists of the senses do not notice such counter evidence is due to the situation that they conflate vision with visualization and objectification: other cultures do not "see" if they do not "visualize" things the way modern Western cultures do. What anthropologists of the senses are not aware of is that objectification has not been the result of the dominant position of seeing in the hierarchy of the senses in Western culture, but of a modern epistemology that reduces seeing to visualization. Moreover, anthropologists of the senses follow the logic of a Cartesian rupture between body and mind that considers "the organs of sense as gateways

4. Jonathan Sterne uses the term "the audio-visual litany" for this set of ideas (Sterne 2003: 15).

5. Jonathan Sterne has labeled this as the "zero-sum game" of sensory modalities, "where the dominance of one sense" in a particular culture or historical episode "by necessity leads to the decline of another sense" (Sterne 2003: 16).

between an external, physical world and an internal world of mind" (Ingold 2000: 257). In their approach, culture is the key to what happens at these gateways: culture defines people's perception of, for instance, light and sound. In turn, people's metaphors of sensory experiences express a particular culture's core values. "I see what you mean" would then express the dominant position of seeing in the sensory map of the West. Ingold, however, claims that people only use metaphors to help others understand their experiences and to establish unison. This does not mean that he would not be interested to know how "sensory metaphors are mobilized in discourse" (Ingold 2000: 285–86). But he does not believe in the existence of an interface between sensory organs and mind. Drawing on work by Hans Jonas, James Gibson, and Maurice Merleau-Ponty, he considers sound and light the experiential qualities of an ongoing and two-way process of "engagement between the perceiver and his or her environment" (257, 268). What is more, the "eyes and ears should not be understood as separate keyboards for the registration of sensation but as organs of the body as a whole, in whose movement, within an environment, the activity of perception consists" (268). Setting out from that position, hearing is a form of seeing and seeing a form of hearing, if only because moving our head is crucial in both activities. Vision is, for instance, much more than just taking a snapshot of the environment—it is more like scanning the environment, as happens in the process of hearing. This makes hearing and seeing comparable phenomena, even though this does not imply that they are equivalent.

Howes responded to Ingold by arguing that the latter's roots in phenomenology—next to other strands of thought such as practice theory and ecological psychology—make him focus on the individual rather than on the social or communal, and assume that his own experiences may guide him in studying those of others. In contrast, students of sensory experience in culture should keep considering the role of indigenous beliefs, rites, and representations in practice-related experiences. In Howes's view, Ingold also mistakenly assumes that "the senses work together in 'synergy,'" thus neglecting potentially conflicting perceptual experiences (Howes 2011: 318). Ingold's counterargument has been, however, that his work is not unconnected to the social, but starts off from a notion of the social that is not confined to cultural values and symbolic representations. He aims to study how humans acquire and cultivate skills of perception through practice and training in particular environments dependent on where they stand in relation to others. "Vision, hearing and the rest are aspects of action—ways of attentively going forth in the world; they are not filters in the conversion of external physical stimuli into internal mental representations" (Ingold 2011b: 325). This is how Ingold sums up his position:

> People do not "make sense" of things by superimposing ready-made sensory meanings "on top" of lived experience, so as to give symbolic shape to the otherwise formless material of raw sensation. They do so, rather, by weaving together, in narrative, strands of experience born of practical, perceptual activity. It is out of this interweaving that meanings emerge. (Ingold 2011b: 326)

In this book, we have tried to follow Ingold in his interest in studying what he calls *lived* experience: sensory experiences in the context of particular practices, in our case the practice of driving cars. We have also taken seriously Ingold's advice to study sensory experience through people's involvement with their environment, as we focus on the history of sound and hearing in and around the artifact of the car. Moreover, we have attempted to take the "multisensory nature of lived experience" into account (Mason and Davis 2009: 589), for example, the interplay between the auditory, visual, and kinesthetic. In fact, our story showed that the significance of enjoying the view from the car did not overshadow the experience of what was audible, but actually underlined the importance of sound and vibrations. Automotive engineers tapping into user studies believed that a smooth drive, less noisy and less shaky, enhanced the cinematic experience of driving. Interestingly, audition and vision, hearing and seeing, seemed to mutually enhance each other in this context. Similarly, the sounds of car radio—after initial fears about its distracting nature had been appeased—were considered to help turn the car into a magic carpet, even though cars increasingly stood still in the traffic jams of the 1960s and after. Music kept the driver moving, if only in his or her dreams. And a few decades later, when noise barriers started blocking the view of drivers, or at least their *control* over what they saw—a situation deeply deplored by drivers since it provoked an "underground" experience—audio entertainment could restore the ideal of control over space, the interior space of the car and its acoustics rather than anything else. Moreover, the modes of listening the driver was supposed to master, to perform as skill, drastically shifted over time: from diagnostic listening to car engines, via distracted listening to car radio, to music listening as a technology of the self for mood control, to, more recently, the ability to switch between such modes of listening and understanding broadcasts of traffic information.

We would not have been able to draft our story, however, without the legacy of the anthropological work that has unraveled the symbolic and cultural dimensions of how people talk about their experiences. We have interpreted the most recent television commercials trying to sell the increasingly subtle sonic differences between cars as musical scores that provide a notation and script for the sonic experiences and related moods the driver is supposed to have when driving the advertised car. Such developments are understandable in a situation in which manufacturers believe we live in an experience society in which products cannot be sold solely through their technical performance anymore, but need stories about what people can experience and feel while driving. We should not forget, however, how important the symbolic dimensions of selling sound have been from the earliest days of car sound design—even though it did not bear that name yet—in the 1920s. Reducing noise was about creating a sense of trust in the car's reliability, but silence was also sold by referring to its convenience and aristocracy. In doing so, advertisers made use of long-standing cultural connotations. They published images of animals that exuded both strength and silence (the swan, the panther) and tapped into the links between wealth and tranquility—controlling one's acoustic environment—or between elites and the ability to keep deferentially silent. At the very same time, they unlinked earlier, or *other*, notions that connected noise to power.

It is important to note, however—and here we may take up a position midway between Ingold and Howes, and close to Pink—that in weaving the narratives of experience, people do not resort only to long-standing symbolic repertoires, but also to their involvement with the technologies of their times. It would not have been possible to liken the experience of driving a car with glass windows to watching a movie if cinema had not yet existed. And if visiting cinemas had not been such a highly valued leisure activity in the 1920s, it would not have been a likely practice for the automotive industry to project it into their cars. As many STS scholars have stressed, such technology-induced vocabularies affect people's stories about and memories of experience. For example, people's near-death experiences have been molded by the theme of seeing a "movie" of one's life, or seeing "panoramic" images in the years prior to the rise of cinema (Draaisma 2001: 244–45). In our view, taking past experiences of technology, literary tropes, and symbolic traditions into account when writing about technological change contributes both to a truly cultural history of technology and to a material cultural analysis of shifting sensory experiences.

No doubt, Ingold and Howes would both agree with Michael Bull's opening statement in the first issue of the academic journal *Senses & Society* that we should not leave the senses to cognitive psychologists alone (Bull 2007). Indeed, we tried to make clear that focusing too much on psychological explanations for acoustic cocooning draws attention away from how we arrived in a situation of seeking sonic relief in the car. We did not clear psychology from our stories, however. At times, we consulted reports written by psychologists from the decades we studied to have access to the reported experience of contemporary drivers. Such reports have been most helpful when researchers had phrased "open questions" and had recorded quotes of the answers by drivers. And in some cases it has been worthwhile to compare the notions of perception psychologists—for example, their take on the tunnel experience when driving along noise barriers—with the self-reported experiences by drivers, quoted in academic studies or in magazines. Work by psycho-acousticians has also informed us of the relevance of looking into auditory attention, even though this only prompted us to unravel its historical context and articulation.

Scholars doing sensory studies—as well as those studying the perception of sound in the automotive industry, as we have seen in our fifth chapter—underline that they need specific methods to elicit the nearly intangible sensory experiences of everyday life. Jennifer Mason and Katherine Davies, for instance, explained how photo elicitation helped people to express, in rich vocabularies, the experience of family resemblances, such as a daughter with hair that was "a bit mousey": hair in transition from blond to brown (Mason and Davies 2009: 595). At the very same time, they struggled with "instances of fleeting sensory experiences that have vanished before one has been quite able to put a finger on them." These could only be made tangible through "epiphanal" stories about the moment people perceived a particular resemblance, as well as about the experience of perception becoming unavailable afterwards (Mason and Davies 2009: 596–97).

If eliciting everyday sensory experiences is so hard, how then were we able to acquire access to such experiences in the past—apart from collecting stories and reports about drivers' experiences? Elsewhere, one of us has stressed that studying technologies that breach cultural conventions, as in the "breaching experiments" by ethnomethodologist Harold Garfinkel, is highly useful to get access to the taken-for-granted aspects of Western culture. This was helpful, for instance, when studying the introduction of unconventional and radically new musical instruments in pop and classical music cultures (Pinch and Bijsterveld 2003). Historian of technology Joy Parr took this approach when she studied and interviewed Canadians who had experienced major changes in their sensorial environment due to megaprojects like the building of a dam or nuclear plant in or near their hometowns (Parr 2010). We have tried to capture producers', users', and engineers' ideas of and experiences with the car and its acoustic makeup by closely studying archival evidence of the first encounter with a new technology, as we did when we unraveled the early history of car radio. We also learned, however, that a new technology may significantly affect people's experiences only when it has become sufficiently ubiquitous, as in the history of the noise barrier. Similarly, we were able to capture shifts in preferred modes of listening, and thus in training drivers in their use of particular sensory modalities, when focusing on historical moments in which people defended their professional skills, as the story of the German mechanics showed. This may even hold true for the story of testing the sound of car prior to the composition and marketing of target sounds: here the difference between expert and lay listening was at issue.

In this way, studying the cultural practices of technologies-in-use and of technologies-for-sale helped us explain how we arrived at the bizarre situation of finding tranquility and sonic relief in a vehicle initially (in)famous for its noise. It also helped us do justice to the anthropology of the senses, sensory ethnography, and the study of perception of the environment, while using the analytical clash of swords between the promoters of these traditions to clarify *how* we answered our question.

THE ECHO OF THE BOOM CAR

Whenever one of us presented a paper in public about acoustic cocooning and car sound design in the years preceding this publication, a member of the audience would always object: "Okay, yes, acoustic cocooning, but what about the 'boom car'?" Is the boom car not the exact opposite of acoustic cocooning? Is it not a more significant phenomenon than acoustic cocooning because it affects the lives of so many citizens on a daily basis? Would it not be a good idea to study that phenomenon as well? Usually the rest of the audience would nod in agreement with the commentator or would add the convertible with high-volume speakers as another textbook example of a phenomenon that seemed to gainsay everything what we had posited about sound design and the search for sonic relief in the car.

To be fair, the convertible and the boom car are topics that fully deserve to be studied. Each of these cars would warrant a distinct monograph. Sound in a boom car can, indeed, be seen as the opposite of acoustic cocooning, since it seems to be about dominating the soundscape *outside* the car at least as much as filling it *within* the car (figure 6.1).[6] Indeed, the real boom car enthusiast measures the sound of his stereo in terms of the number of

6. See http://journal.davidbyrne.com/2010/02/021410-valentines-day.html (accessed February 13, 2011).

Figure 6.1
Hertz boom car
Source: http://www.flickr.com/photos/vongerman/7156969837/in/photostream, retrieved March 5, 2013.
Courtesy: B-LOW Concepts, Martin Hillmann.

blocks he can reach and cover acoustically (Gilroy 2001: 96–97). We might add that the convertible has at times been recommended as a genuine "sound machine." A recent magazine ad for a convertible Mercedes-Benz did exactly that: "Start the new Mercedes-Benz SLS AMG supercar and the impressive grumble of the V8 engine throws the true car aficionado into ecstasies. Yet the owners of this bruiser with its folding doors happen to consider another sound equally irresistible: the sound of their Bang & Olufsen audio installation" (Anonymous 2010b). Moreover, while the everyday driver takes audio technology for granted, the boom car owner tinkers enthusiastically with it. Admittedly, the boom car and convertible are widespread phenomena. In the words of the US-based National Alliance against Loud Car Stereo Assault, the "incessant and invasive booming bass of the car stereo is a pestilence that has pervaded our lives and communities beyond any reasonable limits of tolerance."[7]

Even though the boom car is clearly invasive, only a minority of drivers pump up the volume so loud that bystanders cannot escape hearing it. For that reason, we decided to leave the boom car aside in most of our studies and focus on the majority of drivers. At this point, however, we would like to return to the issue of the boom car—to argue that

7. http://www.lowertheboom.org/index.htm (accessed January 13, 2011).

boom car owners are not that different from the acoustic cocooning drivers as one might expect.

First, even drivers who are used to playing loud music may experience their automotive listening space as a cocoon. Being touched by powerful vibrations does not exclude a feeling of intimacy and belonging—context is important here. In *Acoustic Territories*, artist and writer Brandon Labelle refers to past and contemporary studies of Chicano culture in Los Angeles.[8] In the Chicano culture of the 1960s, slowly cruising cars and their ever-playing radios expressed, at least to the middle-class researchers observing those "lowriders," a deviant gang culture. Today, however, they also represent *safety zones* within such gang cultures: car clubs and car competitions "can function as alternatives to gang participation" and be "a positive expression of cultural identity" (Labelle 2011: 152). At a more material level, Labelle underlines how the "interlacing of beats and movements, energy and bodies, may... highlight why cars and music have such a special relationship":

> The automobile, with its musical partner, has become an intensely sonorous machine, realizing a radical potentiality in creating listening experiences with great momentum. The car becomes a generative space that affords the listening body a private relation to music, as an enclosed space, while also granting volume and vibration to the intensity of the drive. It boosts the presence of the beat through its particular acoustics properties. As a physical space, the car is an optimal music machine, accentuating surround sound, bass vibration, and diffusion. (Labelle 2011: 142)

For Labelle, the boom car creates an amplification that adds up to the "kick already embedded in music's promise of movement" (Labelle 2011: 142). In his view, beat and vibration are not opposed to the more enclosing and privatizing aspects of listening to music while driving. On the contrary, these dimensions may strengthen each other through what he calls "auditory latching," or vibratory sensing.

Second, car manufacturers have recently started to advertise the cocooning qualities of convertibles as well, by praising innovations such as the "automatic draught stop" intended to reduce turbulence, or air conditioning with convertible mode that has to ensure that the "temperature remains as constant as the driving pleasure."[9] They have also introduced,

8. An earlier version of Labelle's chapter "Street: Auditory Latching, Cars, and the Dynamics of Vibration" in *Acoustic Territories* (2011) appeared in *Senses & Society*, see Labelle (2008).

9. http://500sec.com/four-seasons-four-passengers/ (accessed September 28, 2012); http://www.bmw.com/com/en/newvehicles/1series/convertible/2011/_shared/pdf/1series_convertible_catalogue.pdf (accessed September 28, 2012).

as we have discussed above, audio equipment that adapts frequency output in response to environmental noise. The convertible is thus more than just a "hangover" from an acoustical past in which making noise to impress others was considered an attractive aspect of the car. Moreover, today's convertible has a different status than the open car had in the previous century. While it was seen as a cheap alternative in the interwar years, it is now a luxury item.

Third, we would like to stress that when drivers of boom cars and convertibles crank up their sound system, they aim to sonically control their environment as much as acoustic "cocooners" do when they try to find sonic relief in their cars. Of course, unlike acoustic cocooning, the type of behavior associated with playing the boom has often been considered uncivilized and overly manly—an association with an impressive history (Bijsterveld 2008). To sociologist Paul Gilroy, driving while black with "boomin'-on-board sound systems" emerges as "an intrepid, scaled-up substitute for the solipsistic world of the personal stereo, a kind of giant armoured bed on wheels that can shout the driver's dwindling claims upon the world into dead public space at ever-increasing volume" (Gilroy 2001: 97).

Apparently, Gilroy views the boom car as the voice of those who otherwise have little societal power. Some past commentators came up with interpretations of the same kind when addressing the phenomenon of noisemaking by young people at large (Bijsterveld 2008). Similarly, Anne Sofie Laegran has described the use of loud car sound systems by a group of youngsters in a Norway village as "a fight for control of public space and territory" (Laegran 2003: 94). In the analysis of communication scholar Jeremy Packer, however, American attempts to curtail the use of "bass-heavy stereo systems" by black drivers through regulation, drawing on the argument "that drivers [are] unable to hear important aural signifiers of impending doom or rescue," have been forms of "racial profiling... in the name of safety" (2008: 22, 226, 225). Interpreted in this way, it is the *regulation* of the boom car that reflects a fight for control of public space. Still, Packer considers playing the "booming bass" a way of "marking out one's presence in advance" (225).

Brandon Labelle has termed playing the boom "sonic leakage," or the "ultimate public broadcast." For him, it "bespeaks attempts at occupation, a sort of territorial claim gaining force through the sub-woofer while also remaining mobile, and potentially beyond arrest." The boom is a "potent weapon for broadcasting in full view the legendary aggression at the heart of rhythm, a signal preceding the potential use of the gun"—a form of sexual aggression even (Labelle 2011: 150, 152, 155). Lowriders' culture may be about "sharing," as it originally stemmed from a situation in which people could not individually afford a car, and about securing "relational ties" (Labelle 2011: 157); it is also an occupy-movement old style, to slightly rephrase his claim. Interestingly, Labelle's interpretation ties in with an earlier definition of playing radios in lower-class neighborhoods. When in the interwar years representatives of the middle classes started to complain about the endless playing of mechanical musical instruments (that is: the gramophone and the radio) in Dutch urban settings, representatives of Socialist and Communist parties replied by redefining "making noise" as "sharing sound": people did not annoy each other with sound but shared sound, something the elite was unable to understand properly (Bijsterveld 2003).

No matter one's normative stance toward driving boom cars, however, it is hard to deny that is it about trying to sonically control the space one is dwelling in. Acknowledging this implies that driving boom cars is less different from acoustic car cocooning than one might think at first "hearing." Moreover, both the more "expressive" and more "intimate" ways of listening to music in the car have had at least one effect in common: musicians and music producers increasingly compose and produce their music with the car in mind. In an article on automotive listening, popular music scholar Justin Williams cites music producer and artist Stewart Copeland, who considers his car an ideal listening environment. "You're in this cocoon where you have a really big sound in an enclosed environment. Then there's the fact that you're driving with scenery moving past. . . . When I record an album, I spend months listening to it in the studio. I listen to it every day going back and forth in my car. I check it out on tiny systems. And then I hear it coming out of the radio, so I know what it sounds like" (Copeland quoted in Williams 2010: 165). This quote does not tell us to what extent Copeland amplifies the big sound he is listening to. It does, however, open up another interesting aspect of the rise of acoustic cocooning: how music itself has been tweaked to the acoustic space of the car.

So far, we have only shown how engineers made both car and car audio technologies respond to and afford the experience of listening while driving. But music itself has also been tailored to the phenomenon of automotive listening. Williams explains that music producers and radio stations use "dynamic range compression which decreases the overall range of dynamics for a given track in order to make the music sound louder without increasing peak amplitude." This makes that type of music more fit to be played in environments that produce competing sounds, like cars. Sony Studios in New York City has "a car speaker system built into the studio as part of their reference speaker configurations." Moreover, even producers of classical music recordings have "shifted their aim from 'concert realism' to an 'abstract soundstage' that considers specific playback spaces rather than aiming to replicate a concert-hall performance" (Williams 2010: 165–66). However, adjusting music to the listening conditions in the car has proved relevant not just for what we might consider comparatively quiet music. Gangsta rapper and music producer Dr. Dre writes his music first and foremost for people playing it in their cars: "I make the shit for people to bump in their cars." He doesn't care about clubs, radio, or television—it's the car where people listen to music "more than anything." So, when he does a mix, "the first thing I do is go down and see how its sounds in the car (Dr. Dre quoted in Williams 2010: 168). Music production in all genres, then, from Bach to "boom," has taken the car seriously as listening space. Car booming does not exclude car cocooning altogether, even though it is of a different kind than the acoustic cocooning that figures in the larger part of our book.

SENSITIZING MOBILITY CULTURE

Even though we do not want to smuggle away the existence of the boom car, we would like to return to the rise of the intimate forms of acoustic cocooning once more. What have been and may be the wider consequences of how drivers found sonic relief in the car? What are, for instance, the consequences for our notions of private and public space? What are the implications for people's notions of safety and societal responsibility on the road? And what does it mean for the ongoing dominance of the car as a means of passenger transport in today's mobility culture?

Several of our sound studies' colleagues, such as Paul du Gay et al. (1997), Michael Bull (2000), and Heike Weber (2008, 2009, 2010) have shown that listening to music in public through headphones—as became common with the acceptance of the Walkman in the 1980s—transgressed earlier boundaries between what was considered to belong to private and public space, and how this raised widespread concern. Until that decade, nonprofessional listening with headphones had largely been considered to belong to the domain of the home. Being involved in such listening on the street, or acoustic cocooning while moving through public space, violated social norms. Walkman users were seen to isolate themselves from public life and to endanger themselves and others by being less conscious of the auditory cues informing them of what was happening around them. Mobile acoustic cocooning in public space thus breached two conventions at the same time: that of being open, at least in principle, to interaction with other citizens when moving through public space, and the responsibility to prevent accidents by being attentive.

Fifty years earlier, the introduction of listening to radio in the car also raised public concern. Here, however, the only issue seemed to be public safety. Commentators on car radio expressed fears that listening to radio while driving might negatively affect the quality of driving. In fact, with each new sound technology in the car, such as mobile phones and "speaking" navigation sets, such debates pop up again, and designers struggle with the number of auditory warning signals they can effectively employ without disturbing the motorist's level of attention to the act of driving (Nass and Harris 2009). In the case of radio, however, manufacturers smartly created a niche for listening while driving in the country instead of in busy urban centers, and even turned the argument against radio upside down: listening to the radio would actually help drivers to stay awake on long and boring stretches and thus *enhance* public safety—an argument accepted by automotive journalists and drivers alike. And when the roads became more crowded, manufacturers added the argument that radio and the music it played assisted drivers in controlling their temper amid rising numbers of potentially impolite fellow drivers.

Hardly anyone raised the issue, however, that private listening on public roads might violate the conventions of communal space, as had been the argument against acoustic cocooning with the help of the Walkman. This is, we think, because the idea of the car as a mobile living room had already been established with the rise of the closed sedan. The only exception to the rule may have been the more recent anxiousness about drivers not hearing the sirens of ambulances or fire engines due to having turned up the volume on their audio equipment. By and large, both the rise of sound design and audio design fostered the cultural legitimacy of auditory cocooning while driving, creating auditory privacy on the public road.

This does not imply that the rise of acoustic cocooning did not affect the boundaries between public and private space. First, car radio had been sold as a way to have a companion on lonely roads or in late evenings when people might feel vulnerable, as we may recall when thinking of the 1930s ad in which a female driver has an imaginary male violin player accompanying her through the night. In this way car radio not only made the all-too-familiar—the boring road, or the daily commute—more inspiriting and new, it also made the unfamiliar—the unknown road, the spooky twilight zone—more familiar, accompanied by programs and music already known to the driver. It was a way of protecting and armoring a private space that was felt to be vulnerable, and thus partially open to what might happen in public space. As Angela Blake has shown in her intriguing study of the rise of citizens band (CB) two-way radio on the Los Angeles freeways in the mid-1970s, CB radio became especially popular among lower-middle-class white male drivers who feared they might become the victim of black crime and violence on freeways "defined by an inattentive and unfocused seeing" (Blake 2010: 161).

> With their fellow citizens sealed off behind their own car windows, their eyes staring at the road ahead, and their ears perhaps "tuned" to a music radio station, motorists in distress had little means of attracting reliable help on the freeway. (Blake 2010: 169)

CB radio allowed drivers to contact other CB radio owners, a "male network of 'good buddies'" when they spotted persons displaying suspicious behavior. In this way, they mimicked the police. CB radio in the car thus helped people to cross a putatively dangerous public space and create an "audible sense of order" (Blake 2010: 160). Again, the privacy of the car was felt to be under threat—its boundaries with public space crossed—and crossing the line between the car interior and the car exterior once again through controlled communication with a safer world outside happened to be the answer.

Second, people involved in acoustic cocooning in public space, like Walkman users or those playing music on MP3 devices, report that they feel less visible by others, as Michael Bull (2000, 2007) has documented. Blocking out unwanted sounds, not hearing the noises produced by fellow citizens, and playing one's desired music thus has the remarkable effect of feeling oneself "out of view"—a paradoxical auditory retreat from other people's views. Drivers have mentioned similar experiences: feeling oneself safe from other people's views, at least on the highway, when dwelling in one's private sonic bubble. Drivers experience "their space as one that is hermetically sealed" (Bull 2003: 369). This imaginative experience again thwarts traditional distinctions of private and public space. It seems to be the offspring of the somewhat distancing cinematic experience reported by motorists in the 1920s when they drove cars with glass windows and closed bodies—an experience, as we have seen, that was supported by the enhanced sonic qualities of the car and its audio systems.

These phenomena probably have three less positive societal effects, although we admittedly enter the realm of speculation here. Due to the increasing levels of auditory driving comfort our modern cars afford—less noise, less vibration—we may be less aware of the dazzling speed we drive at than when, for instance, we are observing and hearing cars on a highway from an adjacent parking lot, and not only because of differences in relative speed. This means that the auditory privacy and sonic relief we may now experience in the car, the feeling of being sound and safe, can actually endanger our returning home safe and sound. Our seclusion may make us less conscious of the dangerous speed of our mobility. No matter how many safety technologies like airbags and protecting cages we add to our cars, higher speed will always enhance the risk that comes with driving.

In addition, the sound-induced feeling of being invisible—underlined by the growing use of tinted glass and smaller windows—does not heighten our feelings of responsibility toward fellow drivers. If we believe we cannot be observed, we may be less inclined to assist other drivers when necessary, if only due to the infamous bystander effect. As Richard Sennett has argued,

> the actions needed to drive a car, the slight touch on the gas pedal and the break, the flicking of the eyes to and from the rearview mirror, are micro-notions compared to the arduous physical movements involved in driving a horse-drawn coach. Navigating the geography of modern society requires very little physical effort, hence engagement; indeed, as roads become straightened and regularized, the voyager need account less and less for the people and the buildings on the street in order to move, making minute motions in an ever less complex environment. (Sennett 1994: 18)

The experience of effortless movement has only been enhanced by the introduction of car audio, which may encourage the reduced engagement with others on the road identified by Sennett.

Finally, acoustic cocooning as afforded by sound design and audio has helped motorists to cope with a reduction of control over driving—itself the result of traffic jams, road signs, and noise barriers—and thus to accept road congestion and environmental pollution as if these were facts of life. Instead of creating doubts about the "car system" that keeps the carbon-dependent and polluting automobile firmly in its dominant position within our mobility culture (Dennis and Urry 2009: 28), the phenomenon of acoustic cocooning has furnished an imaginative space that keeps us driving as we do.

Acoustic cocooning may even undermine our environmental consciousness in a way that is similar to an effect described by Emily Thompson in the *Soundscape of Modernity* (2002). She explains that acoustical engineers in the first decades of the twentieth century helped control the acoustics of spaces like concert halls, theaters, and offices, and that their increasingly successful attempts to seal these spaces off from street and other city noises made citizens less inclined to fight against the noises and pursue systematic noise abatement. The increasing tranquility of the car interior may have a similar effect. Why bother about traffic noise if you do not hear it in the car?

We detect one particular chance for an opposite trend, though. What if the "acoustic cocooners" among drivers demand the same sonic quality for the environment outside their cars as they do for their cars' interiors? What if they want the artificial tranquility in automotive interiors to be extended to the world beyond their mobile listening booths? In that case, feeling sound and safe in the car would truly sensitize mobility culture. After so many pages full of environmental noise and encapsulated silence, we could not resist ending on this positive note.

AUTHOR BIOGRAPHIES

Karin Bijsterveld is a historian and professor in the Department of Technology and Society Studies, Maastricht University. She is author of *Mechanical Sound: Technology, Culture and Public Problems of Noise in the Twentieth Century* (MIT Press, 2008), and coeditor (with José van Dijck) of *Sound Souvenirs: Audio Technologies, Memory and Cultural Practices* (Amsterdam University Press, 2009). With Trevor Pinch, she has coedited *The Oxford Handbook of Sound Studies* (Oxford University Press, 2012). Bijsterveld is also editor of *Soundscapes of the Urban Past: Staged Sound as Mediated Cultural Heritage* (Transcript Verlag, 2013), and coordinates the research project Sonic Skills: Sound and Listening in Science, Technology and Medicine, for which she was awarded a Vici grant by the Netherlands Organization for Scientific Research.

Eefje Cleophas is a PhD student in the Department of Technology and Society Studies, Maastricht University, with a background in Arts and Sciences. As junior researcher she has been previously involved in the project Complex Interactions Between International Standardization and National Innovation Projects (with Anique Hommels, Tineke Egyedi, and Wiebe Bijker), funded by the Netherlands Organization for Scientific Research (NWO) and the ESF-Eurocores project Europe Goes Critical: The Emergence and Governance of Critical Transnational European Infrastructures. Her current research focuses on the history of standardization processes concerning sound measurement and sound design in the European automotive industry, about which she has, with Karin Bijsterveld, contributed a chapter to *The Oxford Handbook of Sound Studies* (Oxford University Press, 2012).

Stefan Krebs is postdoctoral researcher in the Department of Technology and Society Studies, Maastricht University, and has previously been affiliated with the School of Innovation Sciences at Eindhoven University of Technology. He is author of *Technikwissenschaft als soziale Praxis* (Franz Steiner Verlag, 2008) and has contributed chapters to *The Oxford Handbook of Sound Studies* (Oxford University Press, 2012) and, with Karin Bijsterveld, to *Sonic Interaction Design: Fresh Perspectives on Interactive Sound*, edited by Karmen Franinovic and Stefani Serafin (MIT Press, 2013). He is board member of the German Society for the History of Medicine, Science and Technology (DGGMNT).

Gijs Mom is associate professor in the History of Technology and Mobility at Eindhoven University of Technology. In November 2003, he founded the International Association for the History of Transport, Traffic and Mobility (T^2M). In 2004, Johns Hopkins University Press published his book *The Electric Vehicle: Technology and Expectations in the Automobile Age*. For this book Mom received the ASME Engineer-Historian Award 2004 as well as the Best Book Award from the Society of Automotive Historians. One of his recent publications is "'Historians Bleed Too Much': Recent Trends in the State of the Art in Mobility History," in Peter Norton, Gijs Mom, Liz Millward, and Mathieu Flonneau (eds.), *Mobility in History; Reviews and Reflections (T^2M Yearbook 2012)*, 15–30 (Alphil, 2011). He is currently writing a synthetic book on Atlantic automobilism.

ARCHIVES, JOURNALS, AND INTERVIEWS

ARCHIVES

Algemene Nederlandse Wielrijders Bond (ANWB) (General Dutch Cycling Association), The Hague, the Netherlands. Its current name is Koninklijke Nederlandse Toeristenbond ANWB.

Dansk Standard, Files ISO 362 (1958–1998), Copenhagen, Denmark

Nederlands Akoestisch Genootschap (NAG) (Acoustical Society of the Netherlands), ISO Files, Nieuwegein, the Netherlands

Benson Ford Research Center, Dearborn, MI, United States

Bosch Archives, Stuttgart, Germany

Touring Club de France (TCF), Bibliothèque du 16e Arrondissement, Paris, France

Files Catherine Bertho Lavenir, Paris, France

Daimler Archives, Stuttgart, Germany

Philips Company Archives, Eindhoven, the Netherlands

Personal archives Foort de Roo, former member ISO, TC 43, WG 42, 1993–2001

JOURNALS

ADAC Motorwelt, 1914–33, 1948–59

Allgemeine Automobil-Zeitung, 1908–9, 1919–41

American Motorist, 1914–41

Auto-Anzeiger, 1929

Auto-Technik, 1919–34

AutoCar, 1920–40

Automobil-Revue, 1954–59

Automobil-Werkstatt, 1959

Automobile Engineer, 1950–80

Automobiltechnische Zeitung, 1935–80

Automotive Industries, 1920–40
Bulletin de l'Automobile Club de France, 1923–29
Chilton's Automotive Industries, 1920–40
Club Journal, 1910–30. The full title was *Published every other Saturday by the Automobile Club of America, For Everybody Interested in Motors, Automobiles, Motorboats, Aeronautics, Good Roads, Good Laws, Touring, Travel, Transportation and Club News (New York)*. Its title later changed to *The Automobile Club of America Club Journal*, *The Club Journal: Official Bulletin of the Automobile Club of America*, and *The Club Journal; Published by The Automobile Club of America*.
Das Garagenwesen, 1926–36
De Autokampioen, 1932–40
De Kampioen, 1915–40, 1980–2010
Der Motorwagen, 1898–29. (The years 1901, 1903, 1910–12, and 1920 are not available in the Bosch Archives.)
Dienst am Kunden, 1927–29
FISITA Proceedings, 1960–82
INTERNOISE, 1972–80
Journal de la Société des Ingénieurs d'Automobile 1927–39, 1950–75
Journal of Sound & Vibration, 1964–80
Kampf dem Lärm, 1970–80
Krafthand, 1928–41
La Revue du Touring Club de France, 1920–40
La Technique Automobile et Aérienne, 1919–21, 1926/27
La Vie Automobile, 1920–39
Motor, 1914–31
Motor-Kritik, 1929–34
NOISECON, 1973–81
Noise Control Engineering, 1973–80
PIARC Bulletin, 1951/52, 1960/61, 1970/71, 1980/81, 1990/91
SAE Journal, 1920–45, 1950–80
SAE Transactions, 1955–75
Tankstelle & Garage, 1937/38
Tankstellen- und Garagenbetrieb, 1955–69
Werkstattausgabe der Kfz-Wirtschaft 1949–51

INTERVIEWS

Truls Berge, member of ISO WG 42, and researcher at SINTEF ICT, Department of Acoustics, September 6, 2010, Geneva, Switzerland by Eefje Cleophas (EC)

Gijsjan van Blokland, former member of ISO WG 42, advisor at M+P, and Bart Peeters (M+P), informal interview by Karin Bijsterveld (KB), February 14, 2011, Maastricht, the Netherlands

Nicolas Chouard, acoustic engineer, formerly involved in OBELICS, June 7, 2004, Aachen, Germany, by Kristin Vetter (KV)

David Delcampe, European Commission, DG Environment, October 23, 2007 by Fleur Fragola (FF)

Peter Ehinger, head of acoustic department of Porsche, former member of ISO WG 42, December 10, 2008, Weissach, Germany (EC)

André Fiebig, employee at HEAD Acoustics, July 31, 2008, Kohlscheid, Germany (EC)

Klaus Genuit, engineer, president of HEAD Acoustics, formerly involved in OBELICS, June 3, 2004, Aachen, Germany (KV), and November 18, 2008, Kohlscheid, Germany (EC)

Erik de Graaff, former member of ISO WG 42, employed by M+P, March 24, 2011, Vught, the Netherlands (EC)

Ralph Heinrichs, engineer, Ford employee, June 8, 2004, Cologne, Germany (KV)

Thomas Hempel, engineer, formerly involved in OBELICS, May 26, 2004, Munich, Germany (KV)

Ian Knowles, European Commission, DG Enterprise, October 9, 2007, Brussels, Belgium (FF)

Boudewijn Kortbeek, advisor at the Dutch Ministry of Infrastructure and Environment, March 24, 2011, Vught, the Netherlands (EC)

Winfried Krebber, engineer, formerly involved in OBELICS, May 6, 2004, Aachen, Germany (KV)

Eric Landel and Virginie Maillard, Renault employees, October 16, 2007, Guyancourt, France (FF)

Bernhard Lange, engineer, Opel employee, June 23, 2004, Rüsselsheim, Germany (KV)

Douglas Moore, current convenor of ISO WG 42, employee at General Motors Corporation, September 5, 2010, Geneva, Switzerland (EC)

Leif Nielsen, secretary of ISO WG 42, standardization consultant at Dansk Standard, June 24, 2009, Charlottenlund, Denmark (EC)

Elif Ozcan, researcher at the Department of Industrial Design at Delft University, July 2, 2009, Delft, the Netherlands (EC)

Stephan Paul, (then) PhD student, as researcher located at HEAD acoustics, June 17, 2004, Bonn, Germany (KV)

Foort de Roo, acoustic engineer, former member of ISO WG 42, June 26, 2008, Delft, the Netherlands (EC)

Ulf Sandberg, enigineer, chair of several ISO and CEN committees, October 1, 2007, Brussels, Belgium (FF), and telephone conversation on May 15, 2008 (EC)

Wolfgang Schneider, European Commission, DG Enterprise, October 18, 2007, Brussels, Belgium (FF)

Richard Schumacher, former convener of ISO WG 42, email interview, December 5, 2010 (EC)

Roland Sottek, engineer, formerly involved in OBELICS, May 14, 2004, Aachen, Germany (Transcription in two parts) (KV)

Heinz Steven, member of ISO WG 42, engineer at HS Data Analysis and Consultancy, September 6, 2010, Geneva, Switzerland (EC)

Gerhard Thoma, Manuel Reichle, and Alfred Zeitler, BMW's acoustics department, November 20, 2008, Munich, Germany (EC)

Rob Vermeulen, director of Total Identity (marketing agency), August 7, 2008, Landgraaf, the Netherlands (EC)

Charles Zhang, acoustic engineer at Renault, November 13, 2008, Guyancourt, France (EC)

REFERENCES

ADVERTISEMENTS

Alfa Romeo. 1922. *La Vie Automobile*, July 10, vii.
Audineau. 1925a. *La Vie Automobile*, March 10, vii.
Audineau. 1925b. *La Vie Automobile*, November 25, iii.
Berliet. 1926. *La Vie Automobile*, January 10, xix.
BMW. 1999. *Elsevier*, October 23, 12–14.
Body by Fisher. 1935. *Good Housekeeping*, January 3, page unknown.
Body by Fisher. 1953. *Life Magazine*, October 19, 178.
Body by Fisher. 1958a. *Life Magazine*, March 24, 86.
Body by Fisher. 1958b. *Life Magazine*, July 7, 86.
Body by Fisher. 1967. *Life Magazine*, April 21, 9.
Brampton. 1927. *La Technique Automobile et Aérienne* 18 (1), xiii.
Brennabor. 1929. *Allgemeine Automobil-Zeitung* 30 (43), 31.
Budd. 1932. *Automotive Industries* 66 (8), 109.
Celoron. 1929. *La Vie Automobile*, January 25, xxvi.
Celoron. 1932. *Automotive Industries* 66 (2), 67.
Chandler. 1924. *National Geographic*, April, page unknown.
Chrysler. 1927. *Motor* 15 (1), 11.
Citroën. 1927. *La Vie Automobile*, September 25, i.
Citroën-Phaeton. 1928. *Allgemeine Automobil-Zeitung* 29 (19), 5.
Delage. 1922. *La Vie Automobile*, September 25, xx.
Delage. 1928. *La Vie Automobile*, July 10, ii.
Delage. 1929. *La Vie Automobile*, February 10, inside of cover page.
Felt. 1932. *Automotive Industries* 66 (1), 3.
Ford Focus. 2005. *Dagblad de Limburger*, January 24, B11.
Hoover Steel Ball Company. 1920. *Automotive Industries* 54 (7), 61.
Hotchkiss. 1928. *La Vie Automobile*, January 10, vii.
Hotchkiss. 1930. *La Vie Automobile*, March 10, xv.
Jewett. 1924. *National Geographic*, April, page unknown.
Labourdette. 1925. *Bulletin de l'Automobile Club de France* 3 (6), n.p.
Lincoln. 1926. *Bulletin de l'Automobile Club de France* 4 (10), n.p.
Lincoln. 1927. *Bulletin de l'Automobile Club de France* 4 (6), n.p.
Mahle. 1940. *Allgemeine Automobil-Zeitung* 50 (49), 839.
Manessius. 1925. *La Vie Automobile*, March 25, xii.
Manessius. 1926. *La Vie Automobile*, June 10, xi.

Morse. 1932. *Automotive Industries* 66 (2), 64.
Panhard et Levassor. 1920. *La Vie Automobile*, May 25, n.p.
Panhard et Levassor. 1931. *La Vie Automobile*, October 25, i.
Peugeot. 1928a. *La Vie Automobile*, October 10, xx.
Peugeot. 1928b. *La Vie Automobile*, November 11, xxx.
Peugeot. 1929. *Bulletin de l'Automobile Club de France* 7 (2), xvii.
Plymouth. 1937. *Life Magazine*, August 30, 1.
Pontiac Silver Streak. 1935. *Good Housekeeping*, January 6, 129.
Renault. 1926a. *La Vie Automobile*, March 10, xi.
Renault. 1926b. *La Vie Automobile*, May 10, iii.
Renault. 1928. *Bulletin de l'Automobile Club de France* 6 (10), n.p.
Renault. 1931. *La Vie Automobile*, July 10, vii.
Repusseau. 1926. *La Vie Automobile*, February 10, xvii.
Silentbloc. 1926. *Bulletin de l'Automobile Club de France* 4 (9), n.p.
Timken Axles. 1932a. *Automotive Industries* 66 (5), 47.
Timken Axles. 1932b. *Automotive Industries* 66 (21), 6.
Voisin. 1927. *La Vie Automobile*, May 25, xv.
Weymann. 1924. *La Vie Automobile*, May 25, xxiii.
Weymann. 1926. *La Vie Automobile*, February 25, xxiii.
Weymann. 1929. *La Vie Automobile*, November 25, n.p.
ZF. 1934. *Motor* 23 (12), 19.

OTHER SOURCES

Aangeenbrug, R. T. 1965. Automobile Commuting: A Geographic Analysis of Private Car Use in the Daily Journey to Work in Large Cities. PhD diss., University of Wisconsin.
Aarsen, T. W. H. J. 1985. Geluidwerende constructies bezien van de weg. *Wegen* 59 (12), 411–18.
Abbott, A. 1988. *The System of Professions: An Essay on the Division of Expert Labor*. Chicago: University of Chicago Press.
Akerboom, S. P. 1988. *Het gebruik en effect van (radio)verkeersinformatie: Een schriftelijk vragenlijstonderzoek*. Leiden: Rijksuniversiteit Leiden.
Alberts, G. 2000. Computergeluiden. In G. Alberts and R. van Dael, eds., *Informatica & samenleving*, 7–9. Nijmegen: Katholieke Universiteit Nijmegen.
Alberts, G. 2003. Een halve eeuw computers in Nederland. *Nieuwe Wiskrant* 22, 17–23.
Alberts, W. 1985. Landschappelijke aspecten van geluidwerende constructies. *Wegen* 59 (12), 406–11.
Alexander, J. W. 1938. Auto-radio. *Philips Technisch Tijdschrift*, April 1, 113–19.
Alexandre, A. 1974. Traffic Noise Control in Europe. *Noise Control Engineering* 2 (2), 69–73.
Altman, I. 1976. Privacy: A Conceptual Analysis. *Environment and Behavior* 8 (1), 7–29.
Angus, R. and Harrys, M. 1983. Tuning in on Traffic. *Citizen-Register*, October 23, 1–4.
Anneveld, H. 1996. Als een rat in de val. *Kampioen* 111 (February), 29–32.
Anonymous. 1904. Das "Klopfen" der Automobilmotoren. *Der Motorwagen* 7 (5), 58–59.
Anonymous. 1905. Über das "Klopfen" von Automobilmotoren. *Der Motorwagen* 8 (5), 118–19.
Anonymous. 1916. Der Antrieb von Nocken- und Nebenwellen durch geräuschlose Kette. *Der Motorwagen* 19 (9), 127–28.
Anonymous. 1919. Das Hämmern und Klopfen des Motors. *Allgemeine Automobil-Zeitung* 20 (20), 17–19.

Anonymous. 1920. Les Pannes. *La Vie Automobile* 16, 317–18, 458.
Anonymous. 1921. La carrosserie de la Société des Moteurs Salmson. *La Vie Automobile* 17, xc.
Anonymous. 1924a. Cours élémentaire d'Automobile. Supplement of *La Vie Automobile*.
Anonymous. 1924b. Les carrosseries Weymann. *La Vie Automobile* 20, clxxxiv.
Anonymous. 1924c. Shimmy, Balloons and Air-Cleaners. *Journal of the Society of Automotive Engineers* 15 (December), 482.
Anonymous. 1925a. Les carrosseries Weymann. *La Vie Automobile* 21, 539.
Anonymous. 1925b. La conduite intérieure extra-légère Paul Audineau. *La Vie Automobile* 21, cxxxiv.
Anonymous. 1925c. Impressions That Are an Insult; Horning Gives Cleveland Section Some Thoughts on Riding-Qualities to Mull Over. *Journal of the Society of Automotive Engineers* 16 (4), 392–94.
Anonymous. 1926. Automobilhandel, Automobilreparaturwerkstätten, Garagengewerbe und Publikum. *Allgemeine Automobil-Zeitung* 27 (51), 17–21.
Anonymous. 1928. Braucht man für Automobil-Reparaturen theoretische Kenntnisse? *Die Reparaturwerkstatt* 1 (3), 26–27 and 1 (4), 41–42.
Anonymous. 1929a. Abhorch-Apparat Auto-Doktor. *Allgemeine Automobil-Zeitung* 30 (23), 7.
Anonymous. 1929b. Wo entsteht das Geräusch? *Auto-Anzeiger* 4 (40), 2–3.
Anonymous. 1930a. Riding-Qualities Research: Six Tests of Muscular and Nerve Fatigue Selected as Result of Dr. Moss's Work. *SAE Journal* 26 (1), 99–101.
Anonymous. 1930b. Riding-Comfort Investigations: Valuable Data on Fatigue, Vibration and Shock-Absorbers Presented and Extensively Discussed. *SAE Journal* 26 (2), 137.
Anonymous. 1930c. Can We Get and Measure Riding Comfort? Moss Coordinates the Measurements of Riding-Qualities—Kindl Discusses Shock Absorbers. *SAE Journal* 26 (6), 712–17.
Anonymous. 1930d. Ein idealer Störungssucher. *Das Kraftfahrzeug-Handwerk* 3 (4), 88.
Anonymous. 1931a. Reparaturen-Chaos, und wer hilft heraus? *Allgemeine Automobil-Zeitung* 32 (2), 7.
Anonymous. 1931b. Die andere Seite... *Allgemeine Automobil-Zeitung* 32 (10), 17–18.
Anonymous. 1932. Abnorme Fahrgeräusche und ihre Ursachen. *Das Kraftfahrzeug-Handwerk* 5 (6), 81–82.
Anonymous. 1933. Hände weg. *Allgemeine Automobil-Zeitung* 34 (14), 5.
Anonymous. 1934a. Noise—Engines—Springing at Car Sessions: Debate is Stirred at Noise Symposium. *SAE Journal* 34 (2), 32–40A.
Anonymous. 1934b. Noise Studies Now Important in Design. *SAE Journal (Transactions)* 34 (2), 62.
Anonymous. 1934c. Am Steuer "zieht" es jetzt nicht mehr. *Das Kraftfahrzeug-Handwerk* 7 (3), 41.
Anonymous. 1936. Der Motor hat einen "Ton." *Das Kraftfahrzeug-Handwerk* 9 (11), 330.
Anonymous. 1938. Zu Ende denken! *Das Kraftfahrzeug-Handwerk* 11 (15), 532.
Anonymous. 1939. Es rattert und quietscht. *Das Kraftfahrzeug-Handwerk* 12 (26), 779–80.
Anonymous. 1948. *The Autocar Handbook for the Motorist*. London: Iliffe & Sons Ltd.
Anonymous. 1950. Der Reiseonkel ist wieder da! *ADAC-Motorwelt* 3 (2), 14.
Anonymous. 1955. Auf der Suche nach Geräuschen. *Krafthand* 28 (12), 405.

Anonymous. 1957. Nieuwe ontwikkeling van autoradio. *Philips Mededelingen*, January, A5.
Anonymous. 1967. Ruim 30 jaar geleden verscheen eerste autoradio-ontvangtoestel. *Philips Koerier*, September 2, 10–11.
Anonymous [1969]. *Nu Uw teller nog op nul staat*. [The Hague]: Koninklijke Nederlandsche Automobiel Club.
Anonymous. 1999. Barriers to Noise. *Noise & Vibration Worldwide* 30 (2), 8–10.
Anonymous. 2005. Travel Companion and Traffic Advisor: Blaupunkt for Cars. *Journal of Bosch History* (Supplement 2), 52–59.
Anonymous. 2006. Electric Cars. *The Economist*, July 29, 77.
Anonymous. 2008. Bijzonder geluidsscherm knooppunt de Hogt. *Aluminium* 23 (7), 21.
Anonymous. 2010a. Funbytes. *FD Persoonlijk*, August 21, 21.
Anonymous. 2010b. Levende geschiedenis. *Residence Special* 23 (12), 20.
Attali, J. 1985. *Noise: The Political Economy of Music*. Manchester: Manchester University Press.
Baauw, M. P. 2008. Van icoon naar standaardisatie. *Cement* 60 (1), 56–59.
Baauw, M. P., Zwart, A., and Pontenagel, H. M. J. 2004. Markant scherm in karakteristieke omgeving. *Cement* 56 (2), 21–28.
Bardou, J.-P., Chanaron, J.-J., Fridenson, P., and Laux, J. M. 1982. *The Automobile Revolution: The Impact of an Industry*. Chapel Hill: University of North Carolina Press.
Beecroft, D. 1919. Conditions in the Automotive Industry Abroad. *Journal of the Society of Automotive Engineers* 4 (6), 521–25.
Belasco, W. J. 1979. *Americans on the Road: From Autocamp to Motel, 1910–1945*. Cambridge, MA: MIT Press.
Berger, M. L. 1979. *The Devil Wagon in God's Country: The Automobile and Social Change in Rural America, 1893–1929*. Hamden, CO: Archon Books.
Berger, M. L. 2001. *The Automobile in American History and Culture: A Reference Guide*. Westport, CT: Greenwood Press.
Berk, G. P. 1961. *Goedkoper autorijden: Praktisch handboek voor de moderne automobilist*. Amsterdam: A.J.G. Strengholt Uitgeversmaatschappij N.V.
Bernhard, U. 2002. Specific Development of a Brand Sound. *AVL Engine and Environment*, 103–15. Conference proceedings, no volume or issue number.
Bertho Lavenir, C. 1999. *La roue et le stylo: Comment nous sommes devenus touristes*. Paris: Jacob.
Bethenod, J. 1938. Vérités peut-être utiles à dire. *Journal de la Société des Ingénieurs d'Automobile* 12 (12), 444.
Bijsterveld, K. 2001. The Diabolical Symphony of the Mechanical Age: Technology and Symbolism of Sound in European and North American Noise Abatement Campaigns, 1900–40. *Social Studies of Science* 31 (1), 37–70.
Bijsterveld, K. 2003. "The City of Din": Decibels, Noise and Neighbors in the Netherlands, 1910–1980. *Osiris* 18, 173–93.
Bijsterveld, K. 2006. Listening to Machines: Industrial Noise, Hearing Loss and the Cultural Meaning of Sound. *Interdisciplinary Science Reviews* 31 (4), 323–37.
Bijsterveld, K. 2008. *Mechanical Sound: Technology, Culture and Public Problems of Noise in the Twentieth Century*. Cambridge, MA: MIT Press.
Bijsterveld, K. 2009. *Sonic Skills: Sound and Listening in the Development of Science, Engineering and Medicine, 1920s–now*. Unpublished proposal for the NWO-VICI competition in the Netherlands.

Bijsterveld, K. 2010. Acoustic Cocooning: How the Car Became a Place to Unwind. *Senses & Society* 5 (2), 189–211.

Bijsterveld, K., and Krebs, S. 2013. Listening to the Sounding Objects of the Past: The Case of the Car. In K. Franinovic and S. Serafin, eds., *Sonic Interaction Design*, 3–25. Cambridge, MA: MIT Press.

Billera, D., Parsons, R. D., and Hetrick, S. A. 1997. Good Fences Make Good Neighbors: Highway Noise Barriers and the Built Environment. *Transportation Research Record*, 1601, 55–63.

Binnebesel, C. 2007. *Vom Handwerk zur Industrie: Der PKW-Karosseriebau in Deutschland bis 1939*. PhD diss., Technische Universität Berlin.

Blake, A. M. 2010. An Audible Sense of Order: Race, Fear, and CB Radio on Los Angeles Freeways in the 1970s. In D. Suisman and S. Strasser, eds., *Sound in the Age of Mechanical Reproduction*, 159–78. Philadelphia: University of Pennsylvania Press.

Bock, C. 1937. Über Lyrik und Dramatik des Autofahrens: Besinnliche Bemerkungen zu einem eiligen Thema. *Berliner Tageblatt und Handels-Zeitung* 128 (March 17), page numberers unknown.

Bodin Danielsson, C., and Bodin, L. 2009. Difference in Satisfaction with Office Environment among Employees in Different Office Types. *Journal of Architectural and Planning Research* 26 (3), 241–57.

Boekhorst, J. K. M. te, Coeterier, J. F., and Hoeffnagel, W. J. C. 1986. *Effecten van rijkswegen op de Beleving*. Wageningen: Rijksinstituut voor Onderzoek in de Bos- en Landschapsbouw "De Dorschkamp."

Boomen, T. van den. 1997. Geluidsscherm voor de A16. *Intermediair* 33 (26), June 26, 43.

Borg, K. L. 2007. *Auto Mechanics: Technology and Expertise in Twentieth-Century America*. Baltimore, MD: Johns Hopkins University Press.

Boulandet, R., Lissek, H., Monney, P., Robert, J, and Sauvage, S. 2008. How to Move from Perception to Design: Application to Keystroke Sound. Paper presented at NOISE-CON 2008, Dearborn, MI, July 28–30.

Bourdieu, P. 1977. *Outline of a Theory of Practice*. Cambridge, UK: Cambridge University Press.

Bourdieu, P. 1984. *Distinction: A Social Critique of the Judgement of Taste*. Cambridge, MA: Harvard University Press.

Bourdieu, P. 1989. Social Space and Symbolic Power. *Sociological Theory* 7 (1), 14–25.

Bourdieu, P. 1990. *The Logic of Practice*. Cambridge, MA: Polity Press.

Bourdieu, P. 1999. *Pascalian Meditations*. Stanford, CA: Stanford University Press.

Brand, J. W. 1947. *De automobiel en zijn behandeling*. Rotterdam: Nijgh & van Ditmar N.V.

Brandenburg, G. C., and Swope, A. 1930. Preliminary Study of Riding-Qualities. *SAE Journal* 27 (3), 355–59.

Briefkasten. 1928a. Nr. 36: Tackende Geräusche im Motor. *Allgemeine Automobil-Zeitung* 29 (7), 31.

Briefkasten. 1928b. Nr. 45: Der Motor klopft. *Allgemeine Automobil-Zeitung* 29 (19), 38.

Briefkasten. 1928c. Nr. 93: Schluchzende Geräusche. *Allgemeine Automobil-Zeitung* 29 (19), 38.

Briefkasten. 1928d. Nr. 155: Das laute Getriebe. *Allgemeine Automobil-Zeitung* 29 (8), 29.

Briefkasten. 1929a. Nr. 269: Heulender Wagen. *Allgemeine Automobil-Zeitung* 30 (18), 30.

Briefkasten. 1929b. Nr. 385: Rätselraten. *Allgemeine Automobil-Zeitung* 30 (43), 28.
Briefkasten. 1930. Nr. 808: Es zirpt. *Allgemeine Automobil-Zeitung* 31 (39), 22.
Briefkasten. 1933. Nr. 2673: Der Neuling als Sachverständiger. *Allgemeine Automobil-Zeitung* 34 (39), 18.
Briefkasten. 1938. Nr. 12318: Ein merklich heißes Geräusch auf dem Gaspedal. *Allgemeine Automobil-Zeitung* 39, 999.
Brinks, M. 2011. Hysterisch in de auto als de borden "50" knipperen. *NRC Weekend*, September 10–11, 12–13.
Brown, R. W., and Dickinson, H. C. 1935. Criteria Are Set for Riding Comfort Research; New Instruments Made. *SAE Journal* 37 (2), 20–23.
Brull, C. 1935. Etude des bruits des voitures automobiles. *Journal de la Société des Ingénieurs d'Automobile* 9 (5), 259.
Buckingham, E. 1925a. Transmission Noise and Their Remedies. *Journal of the Society of Automotive Engineers* 17 (July), 62–63.
Buckingham, E. 1925b. Transmission Noise and Their Remedies. *Journal of the Society of Automotive Engineers* 17 (November), 460–62.
Bull, M. 2000. *Sounding out the City: Personal Stereos and the Management of Everyday Life*. New York: Berg.
Bull, M. 2001. Soundscapes of the Car: A Critical Ethnography of Automobile Habitation. In D. Miller, ed., *Car Cultures*, 185–202. New York: Berg.
Bull, M. 2003. Soundscapes of the Car: A Critical Study of Automobile Habitation. In M. Bull and L. Back, eds., *The Auditory Culture Reader*, 357–74. New York: Berg.
Bull, M. 2004. Automobility and the Power of Sound. *Theory, Culture and Society* 21 (4–5), 243–59.
Bull, M. 2007. *Sound Moves: iPod Culture and Urban Experience*. London: Routledge.
Burke, P. 1993. Notes for a Social History of Silence in Early Modern Europe. In P. Burke, *The Art of Conversation*, 123–41. Ithaca, NY: Cornell University Press.
Burkhardt, O. M. 1925. Wheel Shimmying: Its Causes and Cure. *Journal of the Society of Automotive Engineers* 16 (February), 189–91.
Butsch, R. 2000. *The Making of American Audiences: Fom Stage to Television, 1750–1990*. Cambridge, UK: Cambridge University Press.
Callon, M., Méadel, C., and Rabeharisoa, V. 2002. The Economy of Qualities. *Economy and Society* 31 (2), 194–217.
Camlot, J. 2003. Early Talking Books: Spoken Recordings and Recitation Anthologies, 1880–1920. *Book History* 6, 147–73.
Cartoon La Mode. 1927. *Bulletin de l' Automobile Club de France* 5 (1), 11.
Cazalis, L. 1929. Essai de la voiture Nash "Standard." *La Vie Automobile* 25, 105–6.
Charles-Faroux, R. 1922. Notre Concours. *La Vie Automobile* 18, 409–10.
Charles-Faroux, R. 1926. La mort du torpedo. *La Vie Automobile* 22, 25.
Charles-Faroux, R. 1929a. L'Évolution de la Carrosserie moderne. *La Vie Automobile* 25, 350–55.
Charles-Faroux, R. 1929b. Vers le plus grand silence: Le problème de la transmission. *La Vie Automobile* 25, 577–79.
Charles-Faroux, R. 1931. Une nouvelle application du Silentbloc: La Carrosserie déformable Vanvooren. *La Vie Automobile* 27, 89–90.
Charles-Faroux, R. 1932. Commodités d'usage. *La Vie Automobile* 28, 97–99.
Charles-Faroux, R. 1933. Le silence. *La Vie Automobile* 29, 169.
Charles-Faroux, R. 1934. Contre les bruits. *La Vie Automobile* 30, 89–90.
Charles-Faroux, R. 1935. Les carrosseries actuelles. *La Vie Automobile* 31, 99–102.

Charles-Faroux, R. 1937. Pour rendre la voiture plus agréable. *La Vie Automobile* 33, 107–9.

Chauvierre, M. 1926. Essai de la 11 CV Sizaire Frères. *La Vie Automobile* 22, 381–82.

Chauvierre, M. 1932. La mesure du bruit: La cellule photo-electrique et l'automobile. *Journal de la Société des Ingénieurs d'Automobile* 6 (2), 1650.

Clarke, D. 2007. *Driving Women: Fiction and Automobile Culture in Twentieth-Century America*. Baltimore, MD: Johns Hopkins University Press.

Cleophas, E., and Bijsterveld, K. 2012. Selling Sound: Testing, Designing and Marketing Sound in the European Car Industry. In T. Pinch and K. Bijsterveld, eds., *The Oxford Handbook of Sound Studies*, 102–24. Oxford: Oxford University Press.

Cohan, S., and Hark, I. R., eds. 1997. *The Road Movie Book*. New York: Routledge.

Cohn, L. F. 1981. *Highway Noise Barriers*. Washington, DC: Transportation Research Board, National Research Council.

Collins, H. 2001. Tacit Knowledge, Trust and the Q of Sapphire. *Social Studies of Science* 31 (1), 71–85.

Constant, E. W. 1980. *The Origins of the Turbojet Revolution*. Baltimore, MD: Johns Hopkins University Press.

Constant, E. W. 1983. Scientific Theory and Technological Testability: Science Dynameters, and Water Turbines in the 19th Century. *Technology and Culture* 24 (2), 183–98.

Corbin, A. 1986 [1982]. *The Foul and the Fragant: Odor and the French Social Imagination*. Cambridge, MA: Harvard University Press. Originally published as *Le miasme et la jonquille: L'odorat et l'imaginaire social, XVIIIe–XIXe siècles*. Paris: Aubier Montaigne.

Corbin, A. 1995 [1991]. *Time, Desire and Horror: Towards a History of the Senses*. Cambridge, MA: Polity Press. Originally published as *Le Temps, le Désir et l'Horreur*. Paris: Aubier.

Corbin, A. 1999 [1994]. *Village Bells: Sound and Meaning in the Nineteenth-Century French Countryside*. London: Macmillan. Originally published as *Les cloches de la terre: Paysage sonore et culture sensible dans les campagnes au XIXe siècle*. Paris: Albin Michel.

Crane, H. M. 1939. The Car of the Future. *SAE Journal (Transactions)* 44 (4), 141–44.

Critchfield, R. 1937. Modern Automotive Electrical Equipment. *SAE Journal* 41 (2), 358–92.

CROW. 2001. *Richtlijnen geluidbeperkende constructies langs wegen (GCW-2001)*. Ede: CROW.

CROW. 2007. *Richtlijnen geluidbeperkende constructies langs wegen (GCW-2007)*. Ede: CROW.

Danius, S. 2001. The Aesthetics of the Windshield: Proust and the Modernist Rhetoric of Speed. *Modernism/Modernity* 8 (1), 99–126.

Danius, S. 2002. *The Senses of Modernism: Technology, Perception, and Aesthetics*. Ithaca, NY: Cornell University Press.

Davies, W. J. 1994. Noise Barrier. *Environmental Health* 102 (1), 21–25.

Davis, T. 2008. The Rise and Decline of the American Parkway. In C. Mauch and T. Zeller, eds., *The World Beyond the Windshield: Roads and Landscapes in the United States and Europe*, 35–58. Athens: Ohio University Press; Stuttgart: Franz Steiner Verlag.

Debschitz, U. von, and T. von Debschitz, eds. 2010. *Fritz Kahn: Man Machine/Maschine Mensch*. New York: Springer.

Dehue, T. 1995. *Changing the Rules: Psychology in the Netherlands 1900–1985*. Cambridge, UK: Cambridge University Press.

Dembe, A. E. 1996. *Occupation and Disease: How Social Factors Affect the Conception of Work-Related Disorders*. New Haven: Yale University Press.

Dennis, K., and Urry, J. 2009. *After the Car*. Cambridge, MA: Polity Press.

DeNora, T. 2000. *Music in Everyday Life*. Cambridge, UK: Cambridge University Press.

Desportes, M. 2005. *Paysage en mouvement: Transports et perception de l'espace, XVIIe–XXe siècle*. Paris: Gallimard.

Dietl, S. 1931. *Die Fahrtechnik: Ein Hilfs- und Nachschlagewerk mit dem Kraftwagen richtig umgehen und ihn sicher lenken zu können*. Berlin: R. C. Schmidt.

Dijkshoorn, N. 2011. Liefde voor knopjes. *Financieel Dagblad-Persoonlijk*, April 9, 36–38.

Dill. 1936. Der Aerger mit dem Abholtermin. *Allgemeine Automobil-Zeitung* 37 (35), 12.

Dillenburger, H. 1957. *Das praktische Autobuch*. Gütersloh: G. Bertelsmann Verlag.

Dittrich, M. 2001. Sound of Silence. *TPD in 2000—projecten*, 18.

Douglas, S. J. 1999. *Listening In: Radio and the American Imagination, from Amos 'n' Andy and Edward R. Murrow to Wolfman Jack and Howard Stern*. New York: Times Books.

Draaisma, D. 2001. *Waarom het leven sneller gaat als je ouder wordt: Over het autobiografisch geheugen*. Groningen: Historische Uitgeverij.

Duckeck, H. 1973. 40 Jahre Autoradio in Deutschland. *Technikgeschichte* 40, 122–31.

Dumont, P. 1973. *Quai Javel: Quai André Citroën*. Paris: EPA.

Edelmann, H. 1989. *Vom Luxusgut zum Gebrauchsgegenstand: Die Geschichte der Verbreitung von Personenkraftwagen in Deutschland*. Frankfurt am Main: Verband Öffentlicher Verkehrsbetriebe.

Edgerton, D. 2010. Innovation, Technology, or History: What Is the Historiography of Technology About? *Technology and Culture* 51 (3), 680–97.

Ehrenburg, I. 1931 [1929]. *10PK: Het leven der auto's*. Utrecht: W. de Haan.

Ehrenburg, I. 1999 [1929]. *The Life of the Automobile*. London: Serpent's Tail.

Eijbersen, M. 2008. Geluidsschermen: Hoelang nog? *CROW Etcetera* 4 (1), 2.

Ekkelboom, J. 1999. Kunst weert lawaai: Houten geluidsscherm Zutphen. *Het Houtblad*, 11 (2), 40–41.

Elich, J. 2011. Je neus achterna. *NRC Handelsblad* (Weekend), May 28–29, 5.

England, E. C. G. 1930. The Body Problem and Its Solution. *SAE Journal* 27 (1), 69–77.

Enoch, O. 1928. Die Klopfgrenze. *Auto-Technik* 17 (14), 10–11.

Feld, S. 2003. A Rainforest Acoustemology. In M. Bull and L. Back, eds., *The Auditory Culture Reader*, 223–39. New York: Berg.

Feld, S. 2005. Places Sensed, Senses Placed: Toward a Sensuous Epistemology of Environments. In D. Howes, ed., *Empire of the Senses: The Sensual Culture Reader*, 179–91. New York: Berg.

Fesneau, E. 2009. Le marché du "poste à transistors" en France de 1954 à la fin des années 1960. PhD diss., Université Paris 1, Panthéon-Sorbonne.

Fidell, S. 1996. Questing after the Holy Grail of Psychoacoustics . . . Again! *Sound and Vibration* 30 (5), 18–23.

Firestone, F. A. 1926. Technique of Sound Measurement. *Journal of the Society of Automotive Engineers* 19 (5), 461–66.

Fischer, J. 1927. Wie erkennt man die Ursachen der Motorpannen? *Der Dienst am Kunden* 1 (13), 99–101.

Flink, J. J. 1988. *The Automobile Age*. Cambridge, MA: MIT Press.

Flink, J. J. 1992. The Ultimate Status Symbol: The Custom Coachbuilt Car in the Interwar Period. In M. Wachs and M. Crawford, eds., *The Car and the City,* 154–66. Ann Arbor: University of Michigan Press.
Franz, K. 2005. *Tinkering: Consumers Reinvent the Early Automobile.* Philadelphia: University of Pennsylvania Press.
Franzen, T. 1928. European Roads and American Cars. *SAE Journal* 23 (1), 81–82.
Freimann, R. 1993. Das Auto—Klang statt Lärm. In A.-V. Langenmaier, ed., *Der Klang der Dinge,* 45–57. Munich: Verlag Silke Schreiber.
Fricke, N. 2009. Warn- und Alarmsounds im Automobil. In Georg Spehr, ed., *Funktionale Klänge,* 47–64. Bielefeld: Transcript.
Frost, R. 1914. Mending Wall. In R. Frost, *North of Boston,* 11–13. London: D. Nutt.
Ganzevoort, A. W. 1955. *De auto en zijn baas.* The Hague: N.V. Uitgeverij W. van Hoeve.
Gartman, D. 2004. Three Ages of the Automobile: The Cultural Logics of the Car. *Theory, Culture and Society* 21 (4–5), 169–95.
Gay, P. du, Hall, S., Negus, K., Mackay, H., and Janes, L. 1997. *Doing Cultural Studies: The Story of the Sony Walkman.* London: Sage.
Geluidstichting. 1936. *Verslag van het tweede Anti-Lawaai-Congres te Delft op 21 april 1936 georganiseerd door de Koninklijke Nederlandsche Automobiel Club en de Geluidstichting.* Delft: Geluidstichting.
Gilroy, P. 2001. Driving While Black. In D. Miller, ed., *Car Cultures,* 81–104. New York: Berg.
Goodman, D. 2010. Distracted Listening: On Not Making Sound Choices in the 1930s. In D. Suisman and S. Strasser, eds., *Sound in the Age of Mechanical Reproduction,* 15–46. Philadelphia: University of Pennsylvania Press.
Gott, P. G. 1991. *Changing Gears: The Development of the Automotive Transmission.* Warrendale: Society of Automotive Engineers.
Gottfredson, M., et al. 2001. The Ultimate Testing Laboratory: Carmakers Mutate from Heavy Manufacturers to Consumer Goods Companies. *European Business Journal* 13 (2), 66–73.
Goudriaan, J. C., and van Dool, P. H. 1983. *Technische en procedurele aspecten van het oprichten van akoestische afschermingen langs wegen. Deel I: Onderzoek.* [The Hague]: Interdepartmentale Commissie Geluidshinder.
Govindswamy, K., Hartwig, M., Alt, N., and Wolff, K. 2004. Designing Sound to Build Character. *Automotive Engineering International* 112 (2), 172–76.
Graaf, P. de, and Rétrécy, H. de. 1961 [1955]. *Wij en onze auto.* Zwolle: La Rivière and Voorhoeve. Originally published as *Im Auto Zuhause.* Bielefeld: Verlag Delius, Klasing & Co.
Greinert, W.-D. 1994. *The "German System" of Vocational Education: History, Organization, Prospects.* Baden-Baden: Nomos.
H. G. 1999. Slimme systemen om domme weggebruikers. *Wegen* 73 (4), 16–17.
H. G. 2000a. Dynamische reis- en route-informatie: Goed voor de benutting en het automobilistengeluk. *Wegen* 74 (2), 16–19.
H. G. 2000b. Dynamische reis- en route-informatie: De markt wordt aangeslingerd. *Wegen* 74 (4), 12–15.
H. G. 2001. Modulair geluidsscherm is praktischer en goedkoper. *Wegen* 75 (9), 32–34.
Hacker, O. H. 1932. *Panne unterwegs: Ein Hilfsbuch für Kraftfahrer.* Vienna: Steyrermühl.
Hale, J. E. 1923a. Shoeing a Car with Low-Pressure Air. *Journal of the Society of Automotive Engineers* 13 (1), 41–50.
Hale, J. E. 1923b. The Public's and the Car-Builders' Attitude Toward Balloon Tires. *Journal of the Society of Automotive Engineers* 13 (6), 461–65.

Hård, M., and Jamison, A. 2005. *Hubris and Hybrids: A Cultural History of Technology and Science*. New York: Routledge.

Harper, D. 1987. *Working Knowledge: Skill and Community in a Small Shop*. Chicago: University of Chicago Press.

HEADlines. 2003. Newsletter by HEAD Acoustics. Herzogenrath: *HEAD Acoustics*, April.

Heller, A. 1926. Erklärung der Wirkungsweise von Mitteln zum Verhindern des Klopfens in Fahrzeug-Verbrennungs-Maschinen. *Der Motorwagen* 29 (7), 138.

Hellmut, H. H. 1933. Auf langen Fahrten nie mehr allein. *Allgemeine Automobil-Zeitung* 12, 13–14.

Hempel, T. 2001. *Untersuchungen zur Korrelation auditiver und instrumenteller Messergebnisse für die Bewertung von Fahrzeuginnenraumgeräuschen als Grundlage eines Beitrags zur Klassifikation von Hörereignissen*. Munich: Herbert Utz Verlag.

Herrmann, K. L. 1922. Some Causes of Gear-Tooth Errors and Their Detection. *Journal of the Society of Automotive Engineers* 11 (5), 391–97.

Hess, S. P. 1924a. Automobile Riding-Comfort. *Journal of the Society of Automotive Engineers* 15 (1), 82–85.

Hess, S. P. 1924b. Automobile Riding-Comfort. *Journal of the Society of Automotive Engineers* 15 (6), 543–47.

Hess, S. P. 1933. Automobile Riding-Comfort. *SAE Journal* (Annual Meeting Report), 22C–22D.

Hessler, R. 1926. *Der Selbstfahrer: Ein Handbuch zur Führung und Wartung des Kraftwagens*. Leipzig: Hesse & Becker.

Hicks, H. A., and Parker, G. H. 1939. Harshness in the Automobile. *SAE Journal (Transactions)* 44 (1), 1–7.

Hillquist, R. K. 1979. Sound Measurement Standards for Surface Transportation Vehicles. *Noise Control Engineering* 12 (3), 131–33.

Hinlopen, H. 1971. *Uw auto en de electronica*. Deventer: Kluwer.

Hoed, E. den. 2005. SSSSt, een S! Nieuwe Mercedes S-klasse: concertzaal op wielen. *Auto & Motor Techniek* 2, 1–3. Available at http://www.amt.nl/PageFiles/18787/002_2005_9_46_carrosserie.pdf (retrieved September 21, 2012).

Holtwick, B. 1999. "Handwerk," "artisanat," "small business." Zur Formierung des selbständigen Kleinbürgertums im internationalen Vergleich. *Jahrbuch für Wirtschaftsgeschichte* 1, 163–81.

Horning, H. L. 1925. Bearing of Research Department Work on Car Developments. *Journal of the Society of Automotive Engineers* 17 (2), 189–91.

Hort, W. 1929. Der Strassenlärm. *Der Motorwagen* 32 (9), 185–90.

Howes, D., ed. 2005. *Empire of the Senses: The Sensual Culture Reader*. New York: Berg.

Howes, D. 2010a. Response to Sarah Pink. *Social Anthropology* 18 (3), 333–36.

Howes, D. 2010b. Response to Sarah Pink. *Social Anthropology* 18 (3), 338–40.

Howes, D. 2011. Reply to Tim Ingold. *Social Anthropology* 19 (3), 318–22.

Ihde, D. 1976. *Listening and Voice: A Phenomenology of Sound*. Athens: Ohio University Press.

Ingold, T. 2000. *The Perception of the Environment: Essays on Livelihood, Dwelling and Skill*. New York: Routledge.

Ingold, T. 2011a. Worlds of Sense and Sensing the World: A Response to Sarah Pink and David Howes. *Social Anthropology* 19 (3), 313–17.

Ingold, T. 2011b. Reply to David Howes. *Social Anthropology* 19 (3), 323–27.

Interrante, J. A. 1983. *A Movable Feast: The Automobile and the Spatial Transformation of American Culture 1890–1940*. PhD diss., Harvard University, Cambridge, MA.
Illouz, E. 2007. *Cold Intimacies: The Making of Emotional Capitalism*. Cambridge, MA: Polity Press.
Illouz, E. 2009. Emotions, Imagination and Consumption: A New Research Agenda. *Journal of Consumer Culture* 9 (3), 377–413.
Jacklin, H. M. 1936. Human Reactions to Vibration. *SAE Journal* 39 (4), 401–8.
Jackson, D. M. 2003. *Sonic Branding: An Introduction*. Houndmills: Palgrave Macmillan.
Johnson, A. F. 1922. Passenger-Automobile Body-Designing Problems. *Journal of the Society of Automotive Engineers* 8 (4), 306.
Jonny. 1932. [Car doctor cartoon.] *Allgemeine Automobil-Zeitung* 22 (42), 19.
Jonny. 1937. Kunibert, der kluge Kunde: Die Achseinstellung. *Das Kraftfahrzeug-Handwerk* 10 (1), 24.
Joppe, J. 1958. *Prisma Autoboek: Vraagbaak voor de automobilist of voor wie het wil worden*. Utrecht: Het Spectrum.
Kagan, H.-S. 1937. Les méthodes de mesure de bruit et leur application dans la construction automobile. *Journal de la Société des Ingénieurs d'Automobile* 11 (1), 3–14.
Karsemeijer, J. 1997. Blij met de file. *Kampioen* 112 (April), 5.
Kehrt, C. 2006. "Das Fliegen ist immer noch ein gefährliches Spiel": Risiko und Kontrolle der Flugzeugtechnik von 1908 bis 1914. In G. Gebauer, S. Poser, R. Schmidt, and M. Stern, eds., *Kalkuliertes Risiko*, 199–224. Frankfurt am Main: Campus Verlag GmbH.
Kelman, A. Y. 2010. Rethinking the Soundscape: A Critical Genealogy of a Key Term in Sound Studies. *Senses & Society* 5 (2), 212–34.
Kierdorf, B. 1970. *Service-gids automobielen: Autoradio-ontstoring*. Deventer: Kluwer.
Kihlman, T. 1975. Traffic Noise Control in Sweden. *Noise Control Engineering* 5 (3), 124–30.
Kindl, C. H. 1933. New Features in Shock Absorbers with Inertia Control. *SAE Journal (Transactions)* 32 (5), 172–76.
Kipling, R. 1904. Steam Tactics: The Necessitarian. In R. Kipling, *Traffics and Discoveries*, 112–61. Whitefish, MT: Kessinger Publishing's Rare Prints.
Kirschhofer, A. von. 1970. 20 Millionen stöhnen unter Lärm. *Kampf dem Lärm* 17 (4), 91–93.
Klawitter, G., Amann, P. and Dröger, F. 2005. *Autoradios: Praxistipps zu Tunern, Car-Hifi, Zubehör und Multimedia. Geschichte, Gegenwart und Zukunft*. Baden-Baden: Verlag für Technik und Handwerk GmbH.
Kleef, B. van. 2002. Sluipmoordenaar. *Volkskrant Magazine*, June 8, 41.
Kline, R. 2000. *Consumers in the Country: Technology and Social Change in Rural America*. Baltimore: Johns Hopkins University Press.
König, A. 1919. Die experimentelle Psychologie im Dienste des Kraftfahrwesens. *Allgemeine Automobil-Zeitung* 20 (10), 11–13, and 20 (12), 15–17.
König, H. von. 1929. Auto-Philosophie. *Allgemeine Automobil-Zeitung* 30 (13), 17.
König, W. 2000. *Geschichte der Konsumgesellschaft*. Stuttgart: Franz Steiner Verlag.
Kooman, B. 2003. Bekijk Nederland met andere ogen. *Kampioen* 118 (April), 50–51.
Kortbeek, B. J. F., and Boer, E. de. 1987. *Geluidschermen, geluidhinder en visuele hinder*. The Hague: Ministerie van Volkshuisvesting, Ruimtelijke Ordening en Milieubeheer.

Koshar, R. 2002. Germans on the Wheel: Cars and Leisure Travel in Interwar Germany. In R. Koshar, ed., *Histories of Leisure*, 215–30. New York: Berg.

Koshar, R. 2005. Organic Machines: Cars, Drivers, and Nature from Imperial to Nazi Germany. In T. Lekan & T. Zeller, eds., *Germany's Nature: Cultural Landscapes and Environmental History*, 111–39. New Brunswick, NJ: Rutgers University Press.

Kotzen, B., and English, C. 1999. *Environmental Noise Barriers: A Guide to Their Acoustic and Visual Design*. New York: E & FN Spon.

Kotzen, B., and English, C. 2009. *Environmental Noise Barriers: A Guide to Their Acoustic and Visual Design*. 2nd ed. Milton Park: Taylor & Francis.

Kouwenhoven, E. 2002. Autoband moet fluisteren. *Algemeen Dagblad*, May 4, 43.

Kranenburg, A. J. 1985. Geluidwerende constructies, van bouwwerken tot kunstwerken. *Wegen* 59 (12), 419–26.

Kranenburg, A. J., Otto, H., and de Vrede, H. 1987. Zicht op de vormgeving van geluidschermen. *Geluid en omgeving* 10 (September), 118–24.

Kranendonk, F. 1973. *Akoestische eigenschappen van geluidschermen*. The Hague: Interdepartmentale Commissie Geluidshinder.

Krebs, S. 2011. The French Quest for the Silent Car Body: Technology, Comfort and Distinction in the Interwar Period. *Transfers: Interdisciplinary Journal of Mobility Studies* 1 (3), 64–89.

Krebs, S. 2012a. "Sobbing, Whining, Rumbling": Listening to Automobiles as Social Practice. In T. Pinch and K. Bijsterveld, eds., *The Oxford Handbook of Sound Studies*, 79–101. Oxford: Oxford University Press.

Krebs, S. 2012b. Standardizing Car Sound—Integrating Europe? International Traffic Noise Abatement and the Emergence of a European Car Identity, 1950–1975. *History and Technology* 28 (1), 25–47.

Krebs, S. 2013 submitted. "Dial Gauge versus Senses 1-0": German Auto Mechanics and the Introduction of New Diagnostic Equipment, 1950–1980. Submitted to *Technology & Culture*.

Krell, K. 1980. *Handbuch für Lärmschutz an Straßen und Schienenwegen*. Darmstadt: Otto Elsner Verlaggesellschaft.

Kruysse, H. W. 1990. *Een onderzoek naar de beoordelingsdimensies van het wegbeeld: Fase 1*. Leiden: Rijksuniversiteit Leiden, Faculteit Sociale Wetenschappen.

Kümmet, H. 1939. *Meister im Kraftfahrzeughandwerk*. Berlin: Krafthand.

Kümmet, H. 1941. *Lehrling im Kraftfahrzeughandwerk*. Berlin: Krafthand.

Küster, J. 1907. *Chauffeur-Schule: Theoretische Einführung in die Praxis des berufsmäßigen Wagenführens*. Berlin: R. C. Schmidt.

Küster, J. 1919. *Das Automobil und seine Behandlung*. 7th ed. Berlin: R. C. Schmidt.

Labelle, B. 2008. Pump Up the Bass: Rhythm, Cars, and Auditory Scaffolding. *Senses & Society* 3 (2), 187–203.

Labelle, B. 2011. *Acoustic Territories: Sound Culture and Everyday Life*. New York: Continuum.

Labourdette, J.-H. 1972. *Un siècle de carrosserie francaise*. Lausanne: Edita.

Lachmund, J. 1994. *Der abgehorchte Körper: Zur historischen Soziologie der medizinischen Untersuchung*. Opladen: Westdeutscher Verlag.

Lachmund, J. 1999. Making Sense of Sound: Auscultation and Lung Sound Codification in Nineteenth-Century French and German Medicine. *Science, Technology & Human Values* 24 (4), 419–50.

Laegran, A. S. 2003. Escape Vehicles? The Internet and the Automobile in a Local-Global Intersection. In N. Oudshoorn and T. Pinch, eds., *How Users Matter: The Co-construction of Users and Technologies*, 81–100. Cambridge, MA: MIT Press.

Laffert, K.-A. von. 1931. Fern der Welt und doch verbunden. *Allgemeine Automobil-Zeitung* 34 (22.8), 17.

Laignier, H.-G. 1929. L'influence des facteurs psychologiques sur la construction automobile. *Journal de la Société des Ingénieurs d'Automobile* 3 (6), 604.

Lambert, J.-R. 1927. Une opinion sur l'automobile d'avenir. *La Technique Automobile et Aérienne* 18, 56.

Laux, J. 1992. *The European Automobile Industry*. New York: Twayne Publishers.

Lay, W. E., and Fisher, L. C. 1940. Riding Comfort and Cushions. *SAE Journal (Transactions)* 47 (5), 482–96.

Ledertheil, H. 1919. Vom kleinen Wagen und seinen Zukunfts-Aussichten. *Allgemeine Automobil-Zeitung* 20 (24), 15–17.

Leeuwenberg, E., and Boselie, F. 1986. *Een evaluatie van de visuele eigenschappen van drie typen geluidschermen*. Nijmegen: Vakgroep Psychologische Functieleer.

Lefèvre, G. 1923. Essaie de deux 15 HP Chenard et Walcker. *La Vie Automobile* 19, 360–62.

Lemon, B. J. 1925. Glimpses of Balloon-Tire Progress. *Journal of the Society of Automotive Engineers* 16 (2), 172–82.

Lemon, B. J. 1932. Judging Super-Balloon Tires. *SAE Journal (Transactions)* 31, 403–11.

Lennep, D. J. van. 1953. Psychologie van het chaufferen. In J. H. van den Berg and J. Linschoten, eds., *Persoon en wereld: Bijdragen tot de phaenomenologische psychologie,* 155–67. Utrecht: Erven J. Bijleveld.

Lever, C. 1985. *Scherm in beeld: Een onderzoek naar de beleving van geluidwerende voorzieningen door de weggebruiker*. Rotterdam: RBOI, Adviesbureau voor Ruimtelijk Beleid, Ontwikkeling en Inrichting.

Lieshout, M. van. 2006. De dagelijkse file is er voor de broodnodige rust. *De Volkskrant*, June 10, 2.

Lieshout, M. van. 2007. De file als weldaad. *De Volkskrant*, September 27, 15.

Loewe, A. G. von. 1928. Die große Reparaturmisere. *Auto-Technik* 17 (26), 11–14.

Loewe, A. G. von. 1930. Es zieht...! *Allgemeine Automobil-Zeitung* 31 (22), 60.

Loubet, J.-L. 1990. *Les automobiles Peugeot*. Paris: Economica.

Loubet, J.-L. 2001. *Histoire de l'automobile française*. Paris: Éditions du Seuil.

Ludvigsen, K. E. 1995. A Century of Automobile Body Evolution. *Automotive Engineering* 103, 51–59.

Lundin, P. 2008. Mediators of Modernity: Planning Experts and the Making of the "Car-Friendly" City in Europe. In M. Hård and T. J. Misa, eds., *Urban Machinery: Inside Modern European Cities*, 257–79. Cambridge, MA: MIT Press.

Lury, C. 2004. *Branding: The Logos of the Cultural Economy*. London: Routledge.

Lutz, F. K. 1939. Autoradios. *Allgemeine Automobil-Zeitung* 33, 1076–79.

MacKenzie, D. 1989. From Kwajalein to Armageddon? Testing and the Social Construction of Missile Accuracy. In D. Gooding, T. J. Pinch, and S. Schaffer, eds., *The Uses of Experiment: Studies in the National Sciences*, 409–36. Cambridge, UK: Cambridge University Press.

Maekawa, Z.-I. 1968. Noise Reduction by Screens. *Applied Acoustics* 1 (3), 157–73.

Maillard, P. 1932. Les nouvelles Citroen. *La Vie Automobile* 30 (October 10), 519–21.

Maillard, P. 1933. Les carrosseries et le bruit. *La Vie Automobile* 29, 323–26.

Maillard, P. 1935a. Les tendances modernes en carrosserie. *La Vie Automobile* 31, 129–35.

Maillard, P. 1935b. L'Évolution des Carrosseries. *La Vie Automobile* 31, 373–83.

Maillard, P. 1937. Quelques éléments du confort. *La Vie Automobile* 33, 278–80.

Manguel, A. 1999. *Een geschiedenis van het lezen*. Amsterdam: Ambo.

Marks, C., Fischer, R. G., and Stewart, E. E. 1974. Designing Small Cars to Meet Regulations. *SAE Journal* 82 (10), 33–37.
Marks, L. U. 2002. *Touch: Sensuous Theory and Multisensory Media*. Minneapolis: University of Minnesota Press.
Marsch, P., and Collett, P. 1986. *Driving Passion: The Psychology of the Car*. London: Jonathan Cape.
Martini, B. 1922. *Praktische Chauffeur-Schule*. Berlin: R. C. Schmidt.
Martini, B. 1932. *Het handboek der auto -en rijtechniek*. Amsterdam: N. V. Gebr. Graauw's Uitgevers-Mij. en Boekhandel. Translation of *Praktische Kraftfahrschule*. No additional information on original publication available in publication itself.
Martini, B. 1938. *Praktische Kraftfahrkunde: Eine Kraftfahrfibel für jedermann über Störungen, Stegreifreparaturen, Werkzeugkunde, Prüfungsfragen und -antworten*. Berlin: R. C. Schmidt.
Marx, L. 2000 [1964]. *The Machine in the Garden: Technology and the Pastoral Idea in America*. Oxford: Oxford University Press.
Mason, J., and Davies, K. 2009. Coming to Our Senses? A Critical Approach to Sensory Methodology. *Qualitative Research* 9 (5), 587–603.
Mattern, S. 2007. Resonant Texts: Sounds of the American Public Library. *Senses & Society* 2 (3), 277–302.
Matteson, D. W. 1987. *The Auto Radio: A Romantic Genealogy*. Jackson, MI: Thornridge Publishing.
Mauch, C., and Zeller, T. eds. 2008. *The World Beyond the Windshield: Roads and Landscapes in the United States and Europe*. Athens: Ohio University Press; Stuttgart: Franz Steiner Verlag.
Mauss, M. 1936. Les techniques du corps. *Journal de Psychologie Normale et Pathologique* 32 (3–4), 271–93.
Mayer-Sidd, E. 1931. Reparaturen nach schriftlicher Anleitung. *Das Kraftfahrzeug-Handwerk* 4 (3), 57.
McCaffrey, E. 1999 [1989]. "La 628-E8": La voiture, le progrès et la postmodernité. *Cahiers Octave Mirbeau* 6, 122–41. For English translation see [Octave Mirbeau], *Sketches of a Journey; Travels in an early motorcar. From Octave Mirbeau's journal "La 628-E8." With illustrations by Pierre Bonnard*. London: Philip Wilson Publishers, in association with Richard Nathanson.
McDonald, C. 2008a. *A Companion in the Car: The Rise of Car Radio in the United States, 1929–1959*. Unpublished paper, January 27.
McDonald, C. 2008b. Creating a Companion in the Car: Car Radio in America as a Technological Hybrid, 1929–1959. Unpublished paper for the Annual Conference of the Society for the History of Technology, Lisbon, October 11–14.
McIntyre, S. L. 2000. The Failure of Fordism: Reform of the Automobile Repair Industry, 1913–1940. *Technology and Culture* 41 (2), 269–99.
McShane, C. 1994. *Down the Asphalt Path: The Automobile and the American City*. New York: Columbia University Press.
Medina, J. F., and Duffy, M. F. 1998. Standardization vs Globalization: A New Perspective of Brand Strategies. *Journal of Product & Brand Management* 7 (3), 223–43.
Meijer, E. 1993. Kijk uit voor kunst. *Kampioen* 108 (March), 53–56.
Meijer, Mrs. 2000. Dat doet de deur dicht. *Kampioen* 15 (September), 7–8.
Merki, C. M. 2002. *Der holprige Siegeszug des Automobils 1895–1930: Zur Motorisierung des Strassenverkehrs in Frankreich, Deutschland und der Schweiz*. Vienna: Böhlau.

Merriman, P. 2006. "A New Look at the English Landscape": Landscape Architecture, Movement and the Aesthetics of Motorways in Early Postwar Britain. *Cultural Geographies* 13 (1), 78–105.

Merriman, P. 2007. *Driving Spaces: A Cultural-Historical Geography of England's M1 Motorway*. Oxford: Blackwell.

Meyer, H. [1930?]. *Achter het Auto-stuur: Wenken in het belang van rijder en mechanisme*. Amersfoort: Valkenhoff & Co.

Michel, P. 1997. Octave Mirbeau et le concept de modernité. *Cahiers Octave Mirbeau* 4, 11–32.

Ministerie van Verkeer en Waterstaat, Ministerie van Volkshuisvesting, Ruimtelijke Ordening en Milieubeheer. 1990. *Groene geluidbeperkende constructies? Ja, mits*... The Hague: Ministerie van Verkeer en Waterstaat & Ministerie van Volkshuisvesting, Ruimtelijke Ordening en Milieubeheer.

Ministerie van Verkeer en Waterstaat. 2008. *De resultaten van het Innovatieprogramma Geluid*. The Hague: Ministerie van Verkeer en Waterstaat.

Ministerie van Volkshuisvesting, Ruimtelijke Ordening en Milieubeheer. 1989. *Mag het een meterje hoger zijn? Geluidschermen,- hinder en visuele hinder*. The Hague: Ministerie van Volkshuisvesting, Ruimtelijke Ordening en Milieubeheer.

Mody, C. 2005. The Sounds of Science: Listening to Laboratory Practice. *Science, Technology & Human Values* 30 (2), 175–98.

Mom, G. P. A. 1995. *Die Prothetisierung des Autos: Kultur und Technik bei der Bedienung des Automobils*. Presentation, workshop at the Museum für Technik und Arbeit, Mannheim, January 26 and 27.

Mom, G. P. A. 1997. *Geschiedenis van de auto van morgen: Cultuur en techniek van de Elektrische Auto*. Deventer: Kluwer.

Mom, G. P. A. 2004. *The Electric Vehicle: Technology and Expectations in the Automobile Age*. Baltimore, MD: Johns Hopkins University Press.

Mom, G. P. A. 2007. Translating Properties into Functions (and vice versa): Design, User Culture and the Creation of an American and a European Car (1930–1970). *Journal of Design History* 20 (3), 171–81.

Mom, G. P. A. 2008a. *Cultures of Control: Sensorial Struggles in the Automobile, 1920s –1990s*. Paper prepared for the Annual Conference of the Society for the History of Technology, Lisbon, October 11–14.

Mom, G. P. A. 2008b. "The Future is a Shifting Panorama": The Role of Expectations in the History of Mobility. In W. Canzler and G. Schmidt, eds., *Zukünfte des Automobils: Aussichten und Grenzen der autotechnischen Globalisierung*, 31–58. Berlin: Edition sigma.

Mom, G. P. A. 2011. Encapsulating Culture: European Car Travel, 1900–1940. *Journal of Tourism History* 3 (3), 289–307.

Mom, G. P. A. 2013 submitted. Orchestrating Car Technology: Noise, Comfort, and the Construction of the American Closed Automobile, 1917–1940. Submitted to *Technology & Culture*.

Mom, G. P. A., and Filarski, R. 2008. *Van transport naar mobiliteit: De mobiliteitsexplosie (1900–2000)*. Zutphen: Walburg Pers.

Mom, G. P. A., Schot, J., and Staal, P. 2008. Civilizing Motorized Adventure: Automotive Technology, User Culture, and the Dutch Touring Club as Mediator. In A. de la Bruhèze and R. Oldenziel, eds., *Manufacturing Technology: Manufacturing Consumers: The Making of Dutch Consumer Society*, 141–60. Amsterdam: Aksant.

Montag, S. 2011. Lawaai. *NRC Weekend*, June 18 and 19, 37.

Möser, K. 2002. *Geschichte des Autos*. Frankfurt am Main: Campus Verlag.

Möser, K. 2004. "Der Kampf des Automobilisten mit der Maschine"—Eine Skizze der Vermittlung der Autotechnik und des Fahrenlernens im 20. Jahrhundert. In L. Bluma, K. Pichol and W. Weber, eds., *Technikvermittlung und Technikpopularisierung—Historische und didaktische Perspektiven*, 89–102. Münster: Waxmann.

Möser, K. 2009. *Fahren und Fliegen in Frieden und Krieg: Kulturen individueller Mobilitätsmaschinen 1880–1930*. Heidelberg: Verlag Regionalkultur.

Moss, F. A. 1930a. Measurement of Comfort in Automobile Riding. *SAE Journal* 26 (4), 513–21.

Moss, F. A. 1930b. Bodily Steadiness—A Riding-Comfort Index: Discussion of Dr. F. A. Moss's Summer Meeting Paper. *SAE Journal* 27 (1), 111–14.

Moss, F. A. 1932. New Riding-Comfort Research Instruments and Wabblemeter Applications. *SAE Journal (Transactions)* 30 (4), 182–84.

M. S. 1928. Wie findet man Störungsfehler bei Kraftwagen? *Die Reparatur-Werkstatt* 1 (3), 25–26.

Müller, H.-P. 1993. Review of Gerhard Schulze. Die Erlebnisgesellschaft. Kultursoziologie der Gegenwart. Frankfurt am Main: Campus 1992. *Kölner Zeitschrift für Soziologie* 45 (4), 777–80.

Müller, J. 2011. "The Sound of Silence": Von der Unhörbarkeit der Vergangenheit zur Geschichte des Hörens. *Historische Zeitschrift* 292, 1–29.

Muller, W. J. 1930. Developing a Front-Drive Car: Section Meeting and Annual Meeting Paper. *SAE Journal* 26 (6), 753–62.

Nantet, J. 1958. Marcel Proust et la vision cinématographique. *La revue des lettres modernes: Histoire des idées et des littératures* 5, 36–38, 307–13.

Nass, C., and Harris, H. 2009. Audio in the automobile. Paper prepared for the Sound Studies Conference. Maastricht, November 21–22.

Nass, C., Jonsson, I.-M., Harris, H., Reaves, B., Endo, J., Brave, S., and Takayama, L. 2005. Improving Automotive Safety by Pairing Driver Emotion and Car Voice Emotion. Paper prepared for *CHI* 2005, April 2–7, Portland, OR. Available at http://www.stanford.edu/~nass/p1973-nass.pdf (retrieved August 28, 2012).

Nieuwenhuis, P., and Wells, P. 2007. The All-Steel Body as a Cornerstone to the Foundations of the Mass Production Car Industry. *Industrial and Corporate Change* 16 (2), 183–211.

OBELICS. 1999. OBELICS (Objective Evaluation of Interior Car Sound): Unpublished synthesis report.

OECD. 1971. *Urban Traffic Noise: Strategy for an Improved Environment*. Paris: OECD.

OECD. 1995. *Roadside Noise Abatement: Report Prepared by an OECD Scientific Expert Group*. Paris: OECD.

Oldenziel, R., de la Bruhèze, A. A., and Wit, O. de. 2005. Europe's Mediation Junction: Technology and Consumer Society in the 20th Century. *History and Technology* 21 (1), 107–39.

Olyslager, P. 1971. *Auto-ABC: Constructie, werking en behandeling*. Deventer: Kluwer.

Oosterbaan, W. 2010. Het gevaar van geluidloze auto's. *NRC*, December 9, Wetenschap, 12.

Östergren, M., and Juhlin, O. 2006. Car Drivers Using Sound Pryer—Joint Music Listening in Traffic Encounters. In K. O'Hara and B. Brown, eds., *Consuming Music Together: Social and Collaborative Aspects of Music Consumption Technologies*, 173–90. Dordrecht: Springer.

Ostwald, W. 1921. Geräuschlose Zahnräder. *Auto-Technik* 11 (15), 11. Also published in *Allgemeine Automobil-Zeitung* 22 (31), 29.

Ostwald, W. 1922. Selbstzündungs-Klopfen. *Auto-Technik* 11 (3), 5–6.
Ostwald, W. 1923. Autotechnisches Notizbuch: Isolierung der Karosserie vom Rahmen durch Gummibuffer. *Allgemeine Automobil-Zeitung* 24 (10–11), 36.
Otterspeer, W. 2000. *Het bolwerk van de vrijheid: De Leidse universiteit, 1575–1672.* Amsterdam: Bert Bakker.
Otto, H. 1985. Ontwerpen van geluidwerende voorzieningen. *Wegen* 59 (12), 398–405.
Otto, H. 1991. *Geluidwering met beton: Uitgangspunten en basisontwerpen.* 's-Hertogenbosch: Vereniging Nederlandse Cementindustrie.
Paassen, D. van. 2004. Mobiel daten voor forensen. *Intermediair* 48 (November 23), 31.
Packer, J. 2008. *Mobility without Mayhem: Safety, Cars, and Citizenship* Durham, NC: Duke University Press.
Padmos, C. J., de Roo, F., and Niewenhuys, J. W. 1998. Een beetje nieuw geluid voor geluidsschermen. *Wegen* 72 (7), 28–31.
Parr, J. 2010. *Sensing Changes: Technologies, Environments, and the Everyday, 1953–2003.* Vancouver: UBC Press; Seattle: University of Washington Press.
Parzer-Mühlbacher, A. 1926. *Das moderne Automobil: Seine Konstruktion und Behandlung.* Berlin: R. C. Schmidt.
Paton, C. R. 1938. Ride Controls and Calibration. *SAE Journal (Transactions)* 43 (2), August, 313–18.
Pels, D. 2010. TomTom spelt met filemelding voor ANWB. *Trouw*, December 18, Economie, 14.
Peppink, H. J. 1956. *Veredelde rijkunst & op reis met uw auto.* [The Hague]: Koninklijke Nederlandsche Toeristenbond A. N. W. B. Ad. M. C. Stok, Zuid-Hollandse Uitgeversmaatschappij.
Peppink, H. J., and Swanenburg, B. D. 1954. *Auto encyclopedie: Practische vraagbaak voor de automobilist.* Utrecht: Uitgeversmaatschappij W. de Haan N. V.; Gent: Daphne Uitgaven.
Peters, P. 2006. *Time, Innovation and Mobilities.* London: Routledge.
Petit, H. 1921. L'Automatisme en automobile. *La Vie Automobile* 17, 317–19.
Petit, H. 1922a. La voiture utilitaire. *La Vie Automobile* 18, 323–25.
Petit, H. 1922b. La carrosserie Weymann. *La Vie Automobile* 18, 155–58.
Petit, H. 1923a. Essai de la voiture 20 HP Rolls-Royce. *La Vie Automobile* 19, 167–68.
Petit, H. 1923b. Essai d'une 12 HP Panhard. *La Vie Automobile* 19, 365–66.
Petit, H. 1924a. Conduite intérieure ou Torpedo. *La Vie Automobile* 20, 197.
Petit, H. 1924b. Essai d'une voiture Panhard 16 HP sans soupapes. *La Vie Automobile* 20, 71–72.
Petit, H. 1928. Équilibrage et vibrations. *La Vie Automobile* 24, 309–11, 323–25.
Petit, H. 1929. Carrosserie souple ou rigide? *La Vie Automobile* 25, 238–40.
Petit, H. 1934. La voiture automobile actuelle vue par l'usager. *La Vie Automobile* 30, 236–39.
Petit, H. 1935. Les fausses pannes. *La Vie Automobile* 31, 317–20.
Pfetsch, F. 2008. Bargaining and Arguing as Communicative Modes of Strategic, Social, Economic, Political Interaction. In J. Schueler, A. Fickers, and A. Hommels, eds., *Bargaining Norms, Arguing Standards*, 52–65. The Hague: STT.
Picker, J. M. 2003. *Victorian Soundscapes.* Oxford: Oxford University Press.
Pinch, T. 1993. Testing—One, Two, Three... Testing! Towards a Sociology of Testing. *Science, Technology and Human Values* 18 (1), 25–41.
Pinch, T., and Bijsterveld, K. 2003. "Should One Applaud"? Breaches and Boundaries in the Reception of New Technology in Music. *Technology and Culture* 44 (3), 536–59.

Pinch, T., and Bijsterveld, K. eds. 2012. *The Oxford Handbook of Sound Studies*. Oxford: Oxford University Press.
Pink, S. 2009. *Doing Sensory Ethnography*. London: Sage.
Pink, S. 2010a. The Future of Sensory Anthropology/The Anthropology of the Senses. *Social Anthropology* 18 (3), 331–33.
Pink, S. 2010b. Response to David Howes. *Social Anthropology* 18 (3), 336–38.
Plack, C. J., ed. 2010. *The Oxford Handbook of Auditory Science: Hearing*. Oxford: Oxford University Press.
Polanyi, M. 1958. *Personal Knowledge*. London: Routledge & Kegan Paul.
Porcello, T. 2004. Speaking of Sound: Language and the Professionalization of Sound-Recording Engineers. *Social Studies of Science* 34 (5), 733–58.
Porcello, T. 2005. Afterword. In P. D. Greene and T. Porcello, eds., *Wired for Sound: Engineering and Technologies in Sonic Cultures,* 269–81. Middletown, CO: Wesleyan University Press.
Praetorius, I. 1913a. Geräuschlose Zahnketten. *Der Motorwagen* 16 (17), 419–21.
Praetorius, I. 1913b. Geräuschlose Zahnketten. *Der Motorwagen* 16 (18), 449–52.
Praetorius, I. 1913c. Geräuschlose Zahnketten. *Der Motorwagen* 16 (19), 471–72.
Principaux Fournisseurs de l'Automobile. 1928. *La Vie Automobile* 24, n.p.
Principaux Fournisseurs de l'Automobile. 1936. *La Vie Automobile* 32, 495–96.
Proust, M. 2009 [1907]. Motoring Days (Impressions de route en automobile). In J. T. Schnapp, *Speed Limits,* 243–46. Miami Beach, FL: The Wolfsonian—Florida International University.
Prudden, T. M. 1934. Noise Treatment in the Automobile. *SAE Journal (Transactions)* 35 (1), 267–70.
Pursell, C. W. 2010. Technologies as Cultural Practice and Production. *Technology and Culture* 51 (3), 715–22.
R. B. 1924. Neue Wege im Karosseriebau. *Auto-Technik* 13 (4), 20.
Rdl. 1936. Mehr Freude am Basteln. *Allgemeine Automobil-Zeitung* 37 (23), 18–19.
Rdl. 1938. Am Sonntag wurde gebastelt. *Allgemeine Automobil-Zeitung* 39 (38), 1147–49.
Reparatur-Werkstatt. 1929. "Es geht weiter vorwärts!" *Die Reparatur-Werkstatt* 2 (1), 1–2.
Repik, E. P. 2003. Historical Perspective on Vehicle Interior Noise Development. *SAE International,* May, 5.
Rice, T. 2012. Sounding Bodies: Medical Students and the Acquisition of Stethoscopic Perspectives. In T. Pinch and K. Bijsterveld, eds., *The Oxford Handbook of Sound Studies,* 298–319. Oxford: Oxford University Press.
Rice, T., and Coltart, J. 2006. Getting a Sense of Listening: An Anthropological Perspective on Auscultation. *British Journal of Cardiology* 13 (1), 56–57.
Richter, L. 1925. Über das Klopfen der Zündermotoren. *Der Motorwagen* 28 (22), 682–86.
Richter, L. 1926a. Über das Klopfen der Zündermotoren. *Der Motorwagen* 29 (2), 28–32.
Richter, L. 1926b. Über das Klopfen der Zündermotoren. *Der Motorwagen* 29 (13), 281–90.
Richter, L. 1926c. Über das Klopfen der Zündermotoren. *Der Motorwagen* 29 (17), 374–75.
Riesenbeck, H., and Perry, J. 2009. *Power Brands. Measuring, Making, and Managing Brand Success*. Weinheim: Wiley-VHC Verlag.
Rijkswaterstaat. 1979. *De vormgeving van geluidwerende voorzieningen langs wegen*. [Delft]: Rijkswaterstaat.

Rijkswaterstaat. 1986. *Concept-Richtlijnen voor Geluidbeperkende Constructies langs Wegen 1986 (Concept-GCW 1986).* Delft: Rijkswaterstaat.
Rijkswaterstaat. 1989a. *Inventarisatie geluidbeperkende voorzieningen langs rijkswegen.* Delft: Rijkswaterstaat, Hoofdafdeling Milieu, Onderafdeling Advies Geluid en Trillingen.
Rijkswaterstaat. 1989b. *Handleiding Visueel-Ruimtelijke Analyse.* Delft: Rijkswaterstaat, Dienst Verkeerskunde.
Roberts, P. 1976. *Any Color So Long as It's Black: The First Fifty Years of Automobile Advertising.* New York: William Morrow.
Rogers, E. M. 2003 [1962]. *Diffusion of Innovations.* 5th ed. New York: Free Press.
Rolt, L. T. C. 1950. *Horseless Carriage: The Motor-Car in England.* London: Constable.
Rolt, L. T. C. 1964. *Motoring History.* London: Studio Vista.
Roo, J. A. de. 2004. Nederlandse muur. *Kampioen,* July–August, 6.
Ross, C. 2004. Sight, Sound, and Tactics in the American Civil War. In M. M. Smith, ed., *Hearing History: A Reader,* 267–78. Athens: University of Georgia Press.
Roy-Reverzy, E. 1997. La 628-E8 ou la mort du roman. *Cahiers Octave Mirbeau,* 4 (May), 257–66.
Rubery, M. 2008. Play It Again, Sam Weller: New Digital Audiobooks and Old Ways of Reading. *Journal of Victorian Culture* 13, 58–79.
Rubery, M. 2011. *Audiobooks, Literature, and Sound Studies.* New York: Routledge.
Ruppel, E. 1927. *Die Entwicklung der deutschen Personen-Automobil-Industrie und ihre derzeitige Lage.* Berlin: Wittenberge PDM.
Saldern, A. von. 1979. *Mittelstand im "Dritten Reich".* Frankfurt am Main: Campus Verlag.
Samuels, S., and Ancich, E. 2002. Recent Developments in the Design and Performance of Road Traffic Noise Barriers. *Noise & Vibration Worldwide* 33 (1), 16–23.
Sandberg, U. 1993. Noise Emissions of Road Vehicles—Effect of Regulations. Status Report of an I-INCE Working Party. Paper presented at Internoise 93, Leuven, Belgium, August 24–26, 49–50.
Sandberg, U. 2001. Abatement of Traffic, Vehicle, and Tire/Road Noise—the Global Perspective. *Noise Control Engineering Journal* 49 (4), 170–81.
Schafer, R. M. 1994 [1977]. *The Soundscape: Our Sonic Environment and the Tuning of the World.* Rochester, VT: Destiny Books. Originally published as *The Tuning of the World.* New York: Knopf.
Schaffer, S. 1999. Late Victorian Metrology and Its Instrumentation. A Manufactory of Ohms. In M. Biagioli, ed., *The Science Studies Reader,* 457–78. London: Routledge.
Schaffer, S. 2000. Modernity and Metrology. In L. Guzzetti, ed., *Science and Power: The Historical Foundations of Research Policies in Europe,* 71–91. Luxembourg: Office for Official Publications of the European Communities.
Scharff, V. 1991. *Taking the Wheel: Women and the Coming of the Motor Age.* New York: Free Press.
Schenk, N. 2011. De auto komt tot leven. *FD Persoonlijk,* February 5, 38–41.
Schick, A. 1994. Zur Geschichte der Bewertung von Innengeräuschen in Personenwagen. *Zeitschrift für Lärmbekämpfung* 41 (3), 61–68.
Schiffer, M. B. 1991. *The Portable Radio in American Life.* Tucson: University of Arizona Press.
Schipper, F. 2008. *Driving Europe: Building Europe on Roads in the Twentieth Century* Amsterdam: Aksant.
Schivelbusch, W. 1977. *Geschichte der Eisenbahnreise: Zur Industrialisierung von Raum und Zeit im 19. Jahrhundert.* Munich: Hanser.

Schivelbusch, W. 1979. *The Railway Journey: Trains and Travel in the 19th Century*. Oxford: Basil Blackwell.
Schmal, A. 1912. *Ohne Chauffeur: Ein Handbuch für Besitzer von Automobilen und Motorradfahrer*. Vienna: Friedrich Beck.
Schot, J.W. and Lagendijk, V.C. 2008. Technocratic Internationalism in the Interwar Years: Building Europe on Motorways and Electricity networks. *Journal of Modern European History*, 2 (6), 196-216.
Schulte-Fortkamp, B., Genuit, K., and Fiebig, A. 2007. A New Approach for Developing Vehicle Target Sounds. *Sound and Vibration* 41 (10), 12–17.
Schulze, G. 2005 [1992]. *Die Erlebnisgesellschaft*. Frankfurt am Main: Campus Verlag.
Seiler, C. 2008. *Republic of Drivers: A Cultural History of Automobility in America*. Chicago: University of Chicago Press.
Sennett, R. 1994. *Flesh and Stone: The Body and the City in Western Civilization*. New York: Norton.
Sennett, R. 2008. *The Craftsman*. London: Allen Lane.
Sheller, M. 2004. Automotive Emotions. Feeling the Car. *Theory, Culture & Society* 21 (4–5), 221–42.
Shidle, N. G. 1922. Practical Data Gathered for Use in Selling Cars. *Automotive Industries* 45 (8), 351–54.
Shove, E. 1998. Consuming Automobility: A Discussion Paper. Scenarios for a Sustainable Society: Car Transport Systems and the Sociology of Embedded Technologies, Report 1.3. Dublin: Employment Research Centre, Trinity College Dublin. Available at http://www.tcd.ie/ERC/pastprojects/carsdownloads/Consuming%20Automobility.pdf (retrieved May 10, 2012).
Simpson, M. A. 1976. *Noise Barrier Design Handbook*. Washington, DC: Department of Transportation, Federal Highway Administration Office of Research.
Sloot, M. 2002. Nieuwe richtlijnen voor geluidbeperkende constructies langs wegen. *Wegen* 76 (3), 16–18.
Smith, M. M., ed. 2004. *Hearing History: A Reader*. Athens: University of Georgia Press.
Snoek, B. van der. 1988. Een blik achter de schermen. *Kampioen* 103 (March), 52–54.
Snook, C. 1925. Automobile-Noise Measurement. *Journal of the Society of Automotive Engineers* 17 (1), 115–24.
Solomon, L. N. 1954. *A Factorial Study of the Meaning of Complex Auditory Stimuli (Passive Sonar Sounds)*. PhD diss., University of Illinois.
Spoerl, A. 1963a. *Geen angst voor pech onderweg*. Amsterdam: Elsevier. Originally published as *Der Panne an den Kragen*. Munich: R. Piper & Co. Verlag.
Spoerl, A. 1963b. *Uw vriend de auto*. Amsterdam: Elsevier. Originally published as *Mit dem Auto auf du*. Munich: R. Piper & Co. Verlag.
Stadie, A. 1954. Messung des Kraftfahrzeuglärms. *Automobiltechnische Zeitschrift* 56 (5), 131–33.
Stankewitz, B. 2003. Auβengeräuschgesetzgebung als Restriktion beim Sound Design. *DAGA Proceedings*, Aachen, March 18–20, 230–31.
Stanley, F. C. 1926. Causes and Prevention of Squeaking Brakes. *Journal of the Society of Automotive Engineers* 18 (2), 160–62.
Steketee, M. 2006. Hightech autoruit. *Elsevier Thema Auto*, April, 68–69.
Sterne, J. 2003. *The Audible Past: Cultural Origins of Sound Reproduction*. Durham, NC: Duke University Press.
Stockfelt, O. 1994. Cars, Buildings and Soundscapes. In H. Jarviluoma, ed., *Soundscapes: Essays on Vroom and Moo*, 19–38. Tampere: Tampere University.

Storey, B. B., and Godfrey, S. H. 1996. Highway Noise Barriers: 1994 Survey of Practice. *Transportation Research Record* 1523, 107–15.

Strepp, H. 1964. *Veilig autorijden: De kennis, de kunst, de psychologie*. Utrecht: Prisma-boeken.

Suchman, E. A. 1939. Radio Listening and Automobiles. *Journal of Applied Psychology* 3, 148–57.

Sullivan, J. J. 2003. Walls of Fame. *Public Roads* 66 (6), 10–17.

Taub, A. 1930. Powerplant Economics: Piston Displacement versus Horsepower per Dollar. *SAE Journal* 26 (6), 719–26.

Tea, C. A. 1934. Riding-Comfort and Noise Problem Treatment Studied. *SAE Journal (Transactions)* 35 (3), 339–41.

Testor. 1931. Reparaturenchaos—Geburtswehen eines neuen Handwerks. *Allgemeine Automobil-Zeitung* 32 (4), 18.

Thiébault, R. 1963. La lutte contre le bruit des véhicules automobiles et les resultants obtenus. *Journal de la Société des Ingénieurs d'Automobile* 37 (1), 27–40.

Thompson, E. 2002. *The Soundscape of Modernity: Architectural Acoustics, 1900–1933*. Cambridge, MA: MIT Press.

Truax, B., ed. 1978. *The World Soundscape Project's Handbook for Acoustic Ecology*. Vancouver, BC: ARC Publications.

Truax, B. 1984. *Acoustic Communication*. Norwood, NJ: Ablex.

Truax, B. 1996. Soundscape, Acoustic Communication and Environmental Sound Composition. *Contemporary Music Review* 15 (1), 49–65.

Uittenbogerd, G. H. 1997. De verblindende bijwerking van geluidsschermen. *Wegen* 71 (3), 4–7.

Urry, J. 1999. Automobility, Car Culture and Weightless Travel: A Discussion Paper. Available at http://www.comp.lancs.ac.uk/sociology/papers/Urry-Automobility.pdf (retrieved September 24, 2012).

Urry, J. 2000a. Inhabiting the Car. Available at http://www.comps.lancs.ac.uk/sociology/papers/Urry-Inhabiting-the-Car.pdf (retrieved September 24, 2012).

Urry, J. 2000b. *Sociology beyond Societies: Mobilities for the Twenty-First Century*. New York: Routledge.

Vargovick, R. J. 1972. Noise Source Definition—Exterior Passenger Vehicle Noise. *SAE Journal (Transactions)* 80, 1046–51.

Vereniging Nederlandse Cementindustrie. 1991. *Geluidwering met beton*. 's-Hertogenbosch: Vereniging Nederlandse Cementindustrie.

Verheijen, F. 2004. The Wall: Gebouw en geluidsscherm in vloeiend rood vormgegeven. *Civiele Techniek* 59 (3), 8–12.

Verrips, J., and Meyer, B. 2001. Kwaku's Car: The Struggles and Stories of a Ghanaian Long-Distance Taxi-Driver. In D. Miller, ed., *Car Cultures*, 153–84. New York: Berg.

Vetter, K. 2004. Humming, Hissing & Human Values—Sound Evaluation Tests in Car Acoustics as Mediators between Technological Artifacts, Scientific Methods and Culture. Master thesis, European Inter-University Association on Society, Science and Technology.

Vincent, J. G., and Griswold, W. R. 1929. A Cure for Shimmy and Wheel Kick. *SAE Journal* 24 (4), 388–96.

Vincenti, W. 1990. *What Engineers Know and How They Know It*. Baltimore, MD: Johns Hopkins University Press.

Volti, R. 2004. *Cars and Culture: The Life Story of a Technology*. Baltimore, MD: Johns Hopkins University Press.

Vreuls, Paul. 1998. De Huif van Dordrecht. *Kampioen* 113 (November), 34–36.

W. 1938. Vom Umgang mit Meistern. *Allgemeine Automobil-Zeitung* 39 (22), 695–96.
Walkenhorst, P. 1926. Die Handhabung des Autoreparaturgeschäftes. *Auto-Technik* 15 (13), 17–19; 15 (14), 19–21; and 15 (15), 17–18.
Watts, G. R., Crombie, D. H., and Hothersall, D. C. 1994. Acoustic Performance of New Designs of Traffic Noise Barriers: Full Scale Tests. Journal of Sound and Vibration 177 (3), 289–305.
Weber, H. 2008. *Das Versprechen mobiler Freiheit: Zur Kultur- und Technikgeschichte von Kofferradio, Walkman und Handy*. Bielefeld: Transcript Verlag.
Weber, H. 2009. Taking Your Favorite Sound Along: Portable Audio Technologies for Mobile Music Listening. In K. Bijsterveld and J. van Dijck, eds., *Sound Souvenirs: Audio Technologies, Memory, and Cultural Practices*, 69–82. Amsterdam: Amsterdam University Press.
Weber, H. 2010. Head Cocoons: A Sensori-Social History of Earphone Use in West Germany, 1950–2010. *Senses & Society* 5 (3), 339–63.
Wedemeyer, E. A. 1929. Das Klopfen der Motoren. *Der Motorwagen* 32 (16), 334–38.
Weijer, B. van de. 2007. Kalm tuffen of bronstig cruisen. *De Volkskrant*, February 3, 9.
Wenzel, S. 2004. Vom Klang zum Lärm. *Neue Zeitschrift für Musik* 165 (2), 34–37.
Weiss, H. E. 2003. *Chrysler, Ford, Durant and Sloan: Founding Giants of the American Automotive Industry*. Jefferson, NC: McFarland.
Wesler, J. E. 1974. Traffic Noise Legislation. *INTER-NOISE Proceedings*, 461–66.
Wetterauer, A. 2007. *Lust an der Distanz: Die Kunst der Autoreise in der "Frankfurter Zeitung"*. Tübingen: Tübinger Vereinigung für Volkskunde.
Wetzel, C. 1914. Vorschläge für Geräuschverminderung an Lastwagenrädern. *Der Motorwagen* 17 (35), 663–64.
White, R. B. 1991. Body by Fisher: The Closed Car Revolution. *Automobile Quarterly* 29, 46–63.
Wiethaup, H. 1966. Lärmbekämpfung in historischer Sicht. Vorgeschichtliche Zeit—Zeitalter der alten Kulturen usw. *Zentralblatt für Arbeitsmedizin und Arbeitsschutz* 16 (5), 120–24.
Wildervanck, C., and Tertoolen, G. 1997. Autogebruik te sturen? *Verkeerskunde* 48 (1), 18–21.
Wilkens, F. W. 1918a. Anregungen zur Bekämpfung von Geräusch im Automobilbau. *Der Motorwagen* 21 (17), 204–08.
Wilkens, F. W. 1918b. Anregungen zur Bekämpfung von Geräusch im Automobilbau. *Der Motorwagen* 21 (18), 219–22.
Williams, J. A. 2010. "You Never Been on a Ride Like This Befo': Los Angeles, Automotive Listening, and Dr. Dre's 'G-Funk.'" *Popular Music History* 4 (2) 160–76.
Williams, R. 2008 [1990]. *Notes on the Underground: An Essay on Technology, Society, and the Imagination*. Cambridge, MA: MIT Press.
Wilson, A. 1963. *Noise: Final Report*. London: Her Majesty's Stationary Office.
Windecker, C. O. 1937. Der Mann, der alles selber machte. *Das Kraftfahrzeug-Handwerk* 10 (16), 496.
Winkler, H. A. 1972. *Mittelstand, Demokratie und Nationalsozialismus*. Cologne: Kiepenheuer & Witsch.
Winkler, O. 1928. *Automobil-Reparaturen*. Halle an der Saale: Wilhelm Knapp.
Wolf, A. M. 1933. Striking Engineering Progress Revealed in 1933 Cars. *SAE Journal* 31 (1), 1–20.
Wolff, P. 1935. Les applications du caoutchouc dans l'industrie automobile. *Journal de la Société des Ingénieurs d'Automobile* 9 (2), 96–114.

Zeitler, A., and Zeller, P. 2006. *Psychoacoustic Modeling of Sound Attributes*. Proceedings SAE World Congress, April 3–6, Cobo Center, Detroit, MI.
Zeller, T. 2010 [2006]. *Driving Germany: The Landscape of the German Autobahn, 1930–1970*. New York: Berghahn Books.
Ziegler, R. 2007. *The Nothing Machine: The Fiction of Octave Mirbeau*. New York: Rodopi.
Zogbaum, E. 1937. *Unter Motor und Fahrgestell*. Berlin: Krafthand.
Zwaag, M. D. van der, Dijksterhuis, C., Waard, D. de, Mulder, B. L. J. M., Westerink, J. H. D. M., and Brookhuis, K. A. 2012. The Influence of Music on Mood and Performance While Driving. *Ergonomics* 55 (1), 12–22.

INDEX